Knowing Humanity in the Social World

Francis X Remedios • Val Dusek

Knowing Humanity in the Social World

The Path of Steve Fuller's Social Epistemology

palgrave
macmillan

Francis X Remedios
Edmonton, Canada

Val Dusek
University of New Hampshire
Durham, NH, USA

ISBN 978-1-137-37489-9 ISBN 978-1-137-37490-5 (eBook)
https://doi.org/10.1057/978-1-137-37490-5

Library of Congress Control Number: 2017963093

Cover illustration: Pattern adapted from an Indian cotton print produced in the 19th century

Printed on acid-free paper

This Palgrave Macmillan imprint is published by Springer Nature
The registered company is Macmillan Publishers Ltd.
The registered company address is: The Campus, 4 Crinan Street, London, N1 9XW, United Kingdom

Foreword

Knowing Humanity in the Social World is an apt title for a book about my work, especially one focused on the period after my having founded "social epistemology" as an interdisciplinary field in the 1980s and 1990s. For nearly 20 years now, I have been in search of this being, the "human," who is supposed to be the subject of significant epistemic activity. From the outset, I took seriously Michel Foucault's observation in *The Order of Things* that the "human" has always been an uncertain entity, straddling the animal and the divine. But whereas Foucault himself affected a pose of indifference to the future of humanity worthy of a Darwinist—easy come, easy go, from the standpoint of geological time—I have struggled with the problem of what is it about our "humanity" that is worthy of preservation and enhancement. In my most recent writings, I speak of the "value added" of being human, as machines are getting better at matching and excelling at our most highly valued traits. The word "perseverance," now an archaism in English but resonant with such great mid-seventeenth-century thinkers as Hobbes and Milton, captures my sense of quest for the human in a world that in both postmodern theory and capitalist practice incentivizes the erasure of just such an entity from the ontological record.

The book concludes with an interview, which captures something of my spontaneous response to questions. However, the reader may not quite realize the import I place on improvisation as a mark of the human—especially when seen against the backdrop of trying to distinguish humans from animals and machines. I have encouraged the recording of my many talks for quite some time, the bulk of which appear on my website. Indeed, I have accepted so many invitations to speak because of the opportunities they afford me to say things in a new context to a new audience, which in turn provides a crucible for forging new content. I rarely if ever speak from a prepared text—the text is what follows the talk for publication purposes. Of course, I have usually thought and written about the talk's topic for quite some time. But what matters is what I say on the day, which functions as a live experiment, the outcome to which then helps me to decide how to go forward.

Ironically, academic hostility to this *modus operandi* is perhaps greatest in the so-called humanities, where one might have expected a level of human performance that rises above that of a lip-syncing paper reader or Powerpoint presenter. But alas, scholasticism reigns and people are praised for sounding exactly like their papers and only taking questions on exactly what those papers said. No wonder that artificial intelligence researchers think that their machines are close to overtaking humans! The standards of performance demanded of the people still regarded as society's premier knowledge producers—academics—are not as high as we have been led to believe. This observation throws into sharp relief the problem of specifying the "value added" of being human.

I owe a debt of gratitude to Remedios and Dusek for taking the trouble to piece together my various writings to arrive at something that is tractable to an academic reader expecting a more scholastic style of presentation. They correctly identify the general themes that cut across my work, but as the reader will quickly see, the academic uptake of my work is rarely done in terms of the trajectory of my own intellectual project. One recent article, briefly noted by Remedios and Dusek, which does a fine—albeit, in the end quite critical—job of offering an overall assessment of my direction of intellectual travel, is William Lynch's "Social Epistemology Transformed." But generally speaking, my work tends to be contextualized in terms of already existing debates, where my interventions are often difficult to accommodate and hence easily misunderstood, if not outright dismissed. Remedios and Dusek identify those dialectical contexts well, often introducing figures whose positions are relevantly compared to my own, even if I have not introduced them myself (e.g. Jeremy Rifkin's *Algeny*), as well as adjudicate on what counts as fair and unfair criticisms of my positions. I also commend the book's glossary. Although I don't use that much jargon by normal academic standards, I tend to be more allusive in terms of range of references and also sometimes deploy already existing terms and phrases somewhat differently from normal usage.

I'm not ashamed to say that I have been opportunistic—or "kairotic," to use the proper rhetorical term—in piggybacking on larger events to develop my own position. I'm always on the lookout for vehicles for my ideas, and as a tenured professor for most of my career, I've been willing to take risks that others have not had the freedom to do. These include deconstructing Thomas Kuhn, the most influential theorist of science of the second half of the twentieth century, supporting Intelligent Design theory's claim for space in the biology curriculum alongside the modern evolutionary synthesis, or contributing to the philosophical and political maturation of transhumanism, a position that even most postmodernist academics regard as science fictional. I have found all these episodes incredibly fruitful to my thinking, even as they puzzled many onlookers and increased the level of tension surrounding me in the academy.

Admittedly, given this track record, I am not surprised that some people—including friends—see me as primarily an "intellectual provocateur." However, if you read my works carefully, you will see significant continuity amidst the changing circumstances in which I write. Consider Intelligent Design. My entry into the topic was not only via my own strong academic background in history and philosophy of science—a point stressed by Remedios and Dusek—but also my commitment to applying the affirmative action principle for ideas that I introduced in *The Governance of Science*, which was published five years before my serving as an expert witness for the defense in *Kitzmiller v. Dover Area School District*. Moreover, in Chap. 3 of *Science* vs. *Religion?*, the first book I published in the wake of *Kitzmiller*, the reader will see humanity defined in terms of the dualism between animal and machine that frames the argument of *Humanity 2.0*, the first of my trilogy on transhumanism. As it turns out, my peak participation in the controversy surrounding Intelligent Design coincided with my being the UK partner in a European Union Sixth Framework project on the "knowledge politics" of the "converging technologies" agenda in transatlantic science policy.

Perhaps the shrewdest chapter in this book is the one on Gnosticism, which rightly concludes that my stance on the concept is far from coherent. Gnosticism has bothered me from the time I started to grapple with Thomas Kuhn's notorious claim that an "Orwellian" approach to the history of science was necessary to keep the latest scientific recruits motivated to engage in the day-to-day drudgery that is work in the lab or the field. Such a history is airbrushed of the twists and turns, the false leads, the backstabbing, the thwarted alternative lines of research, and so on—the sorts of things that professional historians of science, especially under the influence of social constructivism, have been stressing for the past 40 years. Interestingly, Kuhn appeared to think about what was airbrushed in relatively "cognitivist" terms. He did not seem to have in mind the morally and politically compromising positions in which scientists have been placed (e.g. chemistry in the nineteenth century, atomic and genetic research in the twentieth century), greater knowledge of which might equally cause students to think twice about scientific careers.

In any case, I found Kuhn's claim disturbing because it seemed to imply that these recruits wouldn't stay motivated unless they were told lies. Moreover, given that the "Whiggish" characters of these fake histories were simply technical versions of the ones that were routinely peddled in the popular science literature, two questions of profound ethical import arise: (1) Is knowledge of the real history of science likely to be so destructive of people's faith in science that it can't be made public—not even to science students? (2) If the answer to (1) is yes, then in what sense could someone like Kuhn be said to still believe in the epistemic power—if not overall goodness—of science?

Here Kuhn resembled the twelfth-century philosopher Averroes, whose name is now a by-word for liberalism in the Islamic world, but in his day was regarded as a heretic due to questions raised about his own faith, given his belief that the Qur'an's surface contradictions would prove too challenging to the ordinary believer who wished to take the holy book literally. His solution was the notorious Plato-inspired "double-truth doctrine," whereby scholars would be allowed to pursue the holy book's contradictions without pastoral cares, while the priests would present a version fit for public consumption that would have the effect of maintaining the social order. It is to this mentality that I trace the modern Gnostic preoccupation of keeping science and politics separate, which in turn Kuhn applied to the scientific community itself, resulting in a division of labor between practicing scientists and historians of science—one which Stephen Brush memorably dubbed in the 1970s as making history of science "rated X" for scientists.

What bothered me most about such Gnosticism was its fundamentally anti-democratic and patronizing attitude toward people: that is, its "epistemic asymmetry." It is one thing for "elites"—however one defines them—to believe that they know more about people than the people themselves do. It is quite another for elites to use such knowledge to prevent people from exercising their own judgment. The in-between position that I support would have the elites inform people of their supposedly superior knowledge yet let the people respond as they will, with the consequences made public for all to see. I do not believe that the results of such encounters will necessarily vindicate the elites' worst suspicions about the populace. First of all, people may exercise judgment, err and learn from their mistakes. Moreover, the very same encounters may lead the elites to correct themselves, if the populace turn out either to be correct or to rebound from their errors productively. From my broadly neo-Popperian standpoint, the double-truth doctrine is simply a strategy to make elite knowledge unfalsifiable.

As I had already suggested in *The Intellectual* (2005, pp. 23–24), the double-truth doctrine is alive and well when it comes to the definition of humanity. In particular, racism and creationism are taboo topics in our times, mainly because they strongly challenge the boundary conditions on what it means to be human—the former in terms of our difference from animals, the latter in terms of our difference from God. It's in light of this double taboo that an unholy alliance between skeptical biologists and theistic evolutionists has been forged to inhibit public discussion of the origins and purpose of humanity. It often travels under the banner associated with that darling of "thoughtful" secular thinker, Stephen Jay Gould: "non-overlapping magisteria." The effect—which the social sciences, especially sociology, have yet to take with full seriousness—is that humanity continues to be treated as *sui generis* while lacking any positive account of human uniqueness. Instead we're simply told that we're not reducible to our genes or our ancestry. Recalling theology's periodic penchant for defining God in terms of what the deity is *not,* this "apophatic scientism" is intellectually unsatisfying, to say the least—and perhaps even obfuscatory. However, it plays to a postmodern sensibility that revels in mystery.

The above captures my take on Gnosticism before I seriously engaged with transhumanism. It stressed the side of Gnosticism with which I disagreed, namely, its excessively precautionary attitude to humanity. But there is of course a more proactionary side to Gnosticism, one which Eric Voegelin—rightly, in my mind—associated with the purifying power of violence, especially in the name of returning to fundamentals ("revolutionary" in that sense), which has characterized radical movements in the modern era. While it is easy to imagine Voegelin to be talking about socialist and communist movements, he also means more liberal and libertarian strains that, while officially eschewing the outright killing of people, have historically had little difficulty overturning centuries of traditional order.

The relevant violence was most openly expressed toward non-human nature, which requires "rationalization," "appropriation," "exploitation," and "industrialization." The sort of economic "takeoff" that started in the late eighteenth century which is now said to mark the "Anthropocene" in geological time has been a continuous tale of remaking the planet in humanity's image and likeness, increasingly by people—for example, the Chinese—who don't carry the Abrahamic theological baggage that originally motivated the enterprise.

The preference for the "artificial" over the "natural" that has been the calling card of modernity at many levels is reasonably seen as a bootstrapping strategy to lift humanity out of its fallen state, as defined by its default animal nature, with or without St. Augustine's doctrine of original sin to boot. Common to this line of thought is the Gnostic idea that humanity's full potential is held back by its current embodiment. Transhumanism takes this to the next level while remaining within the materialistic framework that has got us to where we are. The proposal of "morphological freedom" by 2016 third-party US presidential candidate Zoltan Istvan as the fundamental "transhuman right" (i.e. the right of humans to exist in any mode of embodiment they wish—be it animal, cyborg, or digital) suggests that it is not too early to start thinking about modes of governance that can establish justice across an ontologically broader range of human differences than the coverage currently provided by laws that address differences in race, class, gender, and ability. The prospect of radically different instantiations of the human condition serves as a normative proxy for what less materialistic philosophies would call "spiritualization" or "divination."

Remedios and Dusek are right to relate my thinking here to the Russian Cosmist tradition, which has subtly but profoundly focused the activities of "progressive" agents worldwide in the nineteenth and twentieth centuries. Key transhumanist themes of immortalization, resurrection, and interstellar habitation can be found in Nikolai Fedorov and his heretical followers. What switched me toward this more positive valuation of Gnosticism was a rethinking of the Gnostic identification of risking ideas and risking lives. The original historic expression of Gnosticism was as an esoteric cult that required initiates to undergo specific training to access the divine within our being. The modern era has been largely about extending this sensibility while protecting people from the worst consequences of exploring their most ambitious ideas. As Popper liked to stress, the reversible character of both democratic elections and scientific experiments makes them exemplary proving grounds for those ideas. Now, however, we are entering a new historical phase, when the very parameters of a "quality" human existence are now under review. It is becoming increasingly difficult to speak of "protecting" people from the worst consequences of their ideas, when they place less value on their current individual embodiment than on some projected version of themselves, which may be "online" rather than "offline." Indeed, the more we relativize the material conditions under which a "human" existence can occur, the more we shall also have to relativize our sense of what counts as benefits and harms to that existence. In this respect, Gnosticism is gradually being incorporated into our natural attitude toward the secular world.

Coventry, UK Steve Fuller

Acknowledgments

For Francis Remedios, this book brings to fruition his work on Fuller since 2003 after the publication of his book *Legitimizing Scientific Knowledge: An Introduction to Steve Fuller's Social Epistemology*. Francis thanks the Society for Social Studies of Science (4S) organizers for Copenhagen 2012 conference where there was a session on Fuller's Humanity 2.0 and for the 2013 San Diego and 2014 Buenos Aires conferences in which he presented on social epistemology. Francis also thanks Martin Tweedale for an opportunity to discuss parts of this book at a Philosophers' Café in Edmonton, Alberta, and Kim Solez and Herbert Yang, University of Alberta, for opportunities to present parts of this book at their classes.

We thank Steve Fuller for his comments and support. We thank James Collier for a Social Epistemology Review and Reply Collective conference at Virginia Tech, Arlington, in May 2017. Finally, we thank the 4S organizers for the Boston 2017 conference where this book was discussed. We thank Joanna O'Neill and Josh Pitt from Palgrave for their assistance.

Francis dedicates this book to his wife Teresa and his family and his wife's family for their love and support. Val dedicates this book to his daughter Lela and his spouse Christy Hammer for their patience.

Contents

CHAPTER 1

Introduction

Abstract The introduction discusses the need for a book on Fuller's later work since 2000. It characterizes Fuller's work on social epistemology, science studies, evolution controversies, the university, the proactionary approach to technology, and nature. It summarizes the content of the book's chapters.

Keywords Social epistemology • Science studies

Other than Steve Fuller's work, there is no other discussion in current literature of sociology of scientific knowledge (SSK), science and technology studies (STS), sociology of science, philosophy of science, epistemology of science, and analytic social epistemology on the impact of scientific knowledge on humanity. Exploration of the impact is important because scientific knowledge is one of the most powerful forces for improvement of the world and humanity. Scientific knowledge based on nanosciences, synthetic biology, and computer technology is changing humanity to humanity 2.0, which is an emerging object of social science and science policy. Humanity here is used to designate the quality that makes humans distinct from non-humans. Fuller's social epistemology explores the changing social dimensions of both scientific knowledge and humanity as

F.X. Remedios, V. Dusek, *Knowing Humanity in the Social World*,
https://doi.org/10.1057/978-1-137-37490-5_1

well as the impact on each other. Fuller's social epistemology takes seriously that knowledge is produced by agents who may be individually embodied but also collectively embedded in certain specifiable relationships that extend over large chunks of space and time. The need for social epistemology arises from an interdisciplinary gap between philosophy and sociology: philosophical theories of knowledge stress normativity without considering their empirical realizability or political and economic consequences. Philosophers are much better at providing definitions of knowledge (e.g. "justified true belief"). Social theories of knowledge describe empirical conditions, but they do not offer normative considerations.

This book is on the uniqueness of Fuller's social epistemology as described above. This book adds to the discussion in *Legitimizing Scientific Knowledge: An Introduction to Steve Fuller's Social Epistemology* (2003), which focuses on normative issues of how scientific knowledge should be organized and legitimized compared to philosophy of science and SSK and is on Fuller's work up to 2000. The change in Fuller's work from 2000 is the move from epistemology to metaphysics in which Fuller is asking the question: what type of being should the knower be? Our explanation of why Fuller moved from the normative organization of knowledge to ontology of the knower is that Fuller is a social constructivist in which the epistemic agent makes knowledge to act in the world. With the impact of the technosciences such as artificial intelligence, synthetic biology, and computer technology on the knower, Fuller's view is that the knower can be enhanced toward transhumanism in which the knower can be disembodied. The transhumanist knower has exceeded the boundaries of human knower, who for the analytic social epistemologist has propositional knowledge and does not make knowledge to act in the world. Fuller views the epistemic agent as a legal person, which includes corporate agency such as the university. With the university as a corporate agent, this book covers Fuller's more recent focus on the university traditionally as the premier site of knowledge production in its struggle with the forces of neoliberalism, which are forces of commercialization.

As an intellectual provocateur, Fuller has taken on Darwinism and defended Intelligent Design (ID). Fuller has taken on Kuhn, upset sociologists and philosophers of science, and defended his version of normative social epistemology. Fuller has taken on science and technology studies (STS) on several issues including post-truth and defended his version of normative social epistemology.

STS is a discipline with which Fuller is most associated. Through the Science Wars in 1990s and 2000s, Fuller defended STS and began the line in his thinking on how science should be best organized compared to how scientists organize science. Though Fuller agrees with STS on social construction of science, Fuller asks how will science be legitimated. In this way, Fuller' social epistemology has been a meta-theory for STS. Fuller's criticism of STS is that STS provides only a descriptive account of science and not a normative account. STS, especially as manifested in the Edinburgh School, demystified the account of scientific knowledge given by philosophers of science. However, Fuller criticizes STS as being a discipline without knowledge policy and as relativistic. In "Is STS all Talk and no Walk?" (2017), which is a recent controversy on post-truth, Fuller challenges STS to see that though STS has unmasked how science is conducted by scientists, STSers do not see that science is a power game (Baker and Oreskes 2017).

Fuller argues that STS as a discipline has failed to develop its own goals and is increasingly client driven in these neoliberal times. With the impact of neoliberalism, a period when clients can strongly influence how academic knowledge is produced, Fuller defends the university, which is a corporate agent, as the premier site for knowledge production for the public good.

It was in *New Frontiers of Science and Technology Studies* (2007) that Fuller defends humanity as a response to STS's discussions initiated by Haraway and Latour, who argue for removal of the distinction of humans and non-humans both in matters of research and of policy with the introduction of humanity 2.0 and amplification of human qualities. With humanity 2.0, Fuller continues his exploration of human enhancement by providing theological arguments of humanity being created in the image and likeness of God and doing science is to participate in the mind of God.

This book's major focus on the changing boundaries conditions as limits of the epistemic agent, the knower, due to the impact of the technosciences. The question is: what kind of being should the knower be? The epistemic agent now is not the same as the knower of the future in which humans can be enhanced through biotechnology, genetic engineering, and synthetic biology (*Humanity 2.0* 2011). *Humanity 2.0* can be considered a milestone in Fuller's later work since it brings together his discussions in other works on the foundation of the social sciences (*New Sociological Imagination* 2006) and the Intelligent Design debate (*Dissent over Descent* 2008). As Fuller's social epistemology is concerned with social transformation of knowledge on humanity's search for transcendence, the exploration of the changing boundary conditions of the epistemic agent is critical. With the enhancement of humanity through biotechnology, genetic engineering, and synthetic biology, the epistemic agent's identity and social epistemic status can change. With advancement of computer and digital technology, avatars can be created and the identity of the epistemic agent is thereby extended. The interface between the epistemic agent and the world has changed because the epistemic agent can be changed through either human enhancement or enhancement to avatars.

With biotechnology, genetic engineering, and synthetic biology that are changing humanity, what does it mean to be human? What is the distinctiveness of humanity? As humanity is the locus of the social sciences, this book focuses on the changing boundary conditions of biology (race) and ideology (religion) for humanity. With the welfare state as the location of the battle between biology and ideology on humanity, Fuller defends the distinctiveness of humanity. Fuller diagnoses the problem of humanity to be a bipolar disorder between our animal nature (biology) and our search for transcendence of nature (ideology). Are we closer to animals as indicated by Darwinism or are we closer to God as indicated by Christianity? In today's terms, the positions can be portrayed to be between the poles of Peter Singer's animal liberation or Ray Kurzweil's spiritual machines.

For Fuller, humanity, which is moral, is the central project of the social sciences that consists of socially organized resistance to the natural selection and natural forces through collective projects such as Christianity, the university, and the state. Participation in large-scale projects allows humans to control or even reverse the effects of natural selection. For Fuller, the classical sociologists Durkheim, Marx, and Weber all concur with his characterization of the project of humanity. Essential to Fuller's concept of the project of humanity is the redistribution of wealth through the state. Fuller recognizes Foucault's notion that the human sciences as a body of knowledge was created in the nineteenth century, and by the twentieth century, man has died—human sciences as a body of knowledge are in question. Fuller connects humanity to transhumanism, which is the view that humanity can be enhanced or redesigned through technology. With converging technologies, which are biotechnology, nanotechnology, and computer technology, humanity can be transformed to an enhanced version of humanity—humanity 2.0.

How did the project of humanity start? Fuller avers that John Duns Scotus started the project of humanity with a univocal theory of predication of God's attributes to man, while Thomas Aquinas has an equivocal or analogical theory of predication of God's attributes to man. Fuller's view is that for Scotus, man's difference to God is by degree, while for Aquinas, man's difference to God is by kind. Humanity is created in the image and likeness of God. For scientists, Bacon, Newton, and Mendel, who are Christians, doing science is participating in the mind of God. With the advance of the nanosciences, biotechechnology, and genetic engineering with which the future of life can be engineered, there have been many voices that science is playing God. From Fuller's perspective, doing science such as nanosciences, biotechechnology, and genetic engineering is to participate of God's mind.

Fuller takes on Darwinianism with Intelligent Design theory. For Fuller, ID is the view of the role of divine design in Western science. In 2005, Fuller was an expert witness to defend ID to be taught in schools at the Kitzmiller versus Dover trial. The judge disagreed that ID is science. Controversially, Fuller, who views humans created in the image and likeness of God, recommends the promotion of theology to motivate students to become scientists in the USA because of theology's view that humans are privileged to understand and control nature. With humanity to participate in the mind of God, humanity as the epistemic agent is imbued with the divine.

The reader may wonder what Intelligent Design and abstract discussions of ancient theology have to do with social epistemology and our place in the world. This makes sense if you see Fuller as a public intellectual as challenging a view of humanity as decentered by the Copernican and Darwinian revolutions, heroically arguing that humans are central to creation because of their ability to construct the world in imitation of the creator. In this Fuller follows the maker's knowledge tradition going back to Bacon and Hobbes and counters Latour's view that we are just another node in networks of things and people. As a public intellectual, Fuller continues to defend humanity with arguments for Intelligent Design in spite of criticisms of his motivations. Fuller's explanation is that though position is unpopular, he has a responsibility as a public intellectual to do so.

Chapter 2 is on Fuller's relation to STS. This chapter starts with a contrast and comparison of Fuller and Latour is included, surveying Fuller's criticisms of Latour (as the leading competitor in science studies), on the human and non-human distinction in which Fuller defends that humans have agency against Latour's view that agency is deflated to actor-network theory. Next is Fuller's treatment of Kuhn and Popper since Fuller views Kuhn as setting the stage for STS's ascendancy through STS criticism of normativity in philosophy of science. Fuller's relation to STS to which Fuller is most closely associated is described. Fuller considers STS to provide a descriptive account of science as opposed to his own normative account. Fuller is critical of STS's lack of knowledge policy and of its relativism. STS's version of social constructivism is contrasted to Fuller's realism on the social sciences and social constructivism on the natural sciences.

With the contrast to Latour's actor-network theory which deflates agency to the networks and actants, Chap. 3 is on agent-oriented social epistemology, which emphasizes epistemic agency or the knower as ontologically open. This is from Fuller's move to transhumanist in which to knower is enhanced to become disembodied. Fuller views the epistemic agent to make knowledge to act in the world as contrasted to analytic social epistemology's epistemic agent, who is a human knower with beliefs that do not make knowledge through construction of reality. Agent-oriented social epistemology, which is about the epistemic agent organizing inquiry, is contrasted to object-oriented social epistemology, which is about the knower believing in objects of knowledge, which represent reality, and not having to construct knowledge to act in the world. There is also a discussion of cognitive economics in which the epistemic agent makes knowledge and leverages beliefs to action instead of the epistemic agent having beliefs to access knowledge.

Chapter 4 is on Fuller's view that the university as a corporate agent is the premier site of knowledge production for the public good. Fuller defends the university against the impact of neoliberalism in which clients influence how academic knowledge is produced. In this context "interdisciplinarity" becomes a battleground. Fuller prefers a version of interdisciplinarity that is regenerative of the university in which academics reach beyond their own fields to a neoliberal version of interdisciplinarity in which academics work in teams for clients on projects. With agent-oriented social epistemology, Fuller's view of interdisciplinarity is that it is internal to agent to organize the disciplines versus object-oriented social epistemology in which disciplines are organized externally by experts. Fuller's view of interdisciplinarity is contrasted to Frodeman's view of transdisciplinarity on whether there are limits to knowledge.

Chapter 5 is on Fuller's version of Intelligent Design (ID) debate with evolutionary theorists and the controversy concerning it. This chapter discusses why ID is important to Fuller in terms of defending the spiritual distinctiveness of humanity, which has epistemic agency, and to do science is to participate in the mind of God since humanity is created in the image and likeness of God. *Fuller's suggestion that Darwin is not a scientist and that evolution is mere natural history in contrast to the proponents of ID theory and molecular biology is criticized, as is Fuller's emphasis on evolution as sheer randomness.*

Chapter 6 is on an exploration of Fuller's version of Cosmism, which is a doctrine originating in early twentieth-century Russia, advocating human resurrection and ultimately human control of the universe, and charges of Gnosticism, which is the Christian heresy holding that the creator of the world was an evil creature and that God was beyond this realm and knowledge (*gnosis*) one can move beyond this world to a higher form of being, against Fuller in terms of transhumanism, which includes epistemic agency. Fuller sees the affinity of his conceptions of transhumanism with Russian Cosmism, growing out of Russian Orthodox theosis, union with God, pursuing space travel, and influencing Teilhard's noosphere. Fuller was called a Gnostic because of the promethean goals of his project, yet his democratic ideal is opposite to Gnosticism. Fuller himself calls biologist opponents of biological race differences of Gnosticism in holding one truth for the masses and another for the scientist elite.

Chapter 7 is on proactionary versus precautionary principle in terms of humanity. The proactionary principle stresses risk-taking as definitive of the human condition, while the precautionary principle stresses the need to conserve nature. Fuller explores the various futures open for the human condition, including ones that embrace "transhumanism" and "posthumanism." Starting with the emerging challenges posed by so-called human enhancement sciences and technologies, Fuller has explored alternative futures under three rubrics: the ecological, the biomedical, and the cybernetic. These attempts to re-engineer both our bodies and the environment require substantial re-definitions of social justice and economics productivity, all envisioned within new political orders of Welfare State 2.0 as opposed to Welfare State 1.0.

Chapter 8 is a Conclusion which is an overall assessment and significance with how Fuller's work has changed from 2000.

The Postscript is an interview with Fuller in 2014 in Buenos Aires at Social Studies of Science conference that involves his reflections on the path his social epistemology has taken to date.

References

Baker, E., & Oreskes, N. (2017). It's No Game: Post-truth and the Obligations of Science Studies. *Social Epistemology Review and Reply Collective, 6*(8), 1–10. http://wp.me/p1Bfg0-3FB

Fuller, S. (2006). *The New Sociological Imagination*. London: Sage Publications, Ltd.

Fuller, S. (2007). *New Frontiers in Science and Technology Studies*. New York: Polity Press.

Fuller, S. (2008). *Dissent on Descent: Intelligent Design's Challenge to Darwinism*. Cambridge: Icon Books.

Fuller, S. (2011). *Humanity 2.0: What It Means to Be Human Past, Present and Future*. London: Palgrave Macmillan.

Fuller, S. (2017). Is STS All Talk and No Walk? *EASST Review, 36*(1) https://easst.net/article/is-sts-all-talk-and-no-walk/

CHAPTER 2

Fuller on Science and Technology Studies

Abstract This chapter starts with a contrast and comparison of Fuller and Latour and a survey of Fuller's criticisms of Latour, on the human and non-human distinction in which Fuller defends that humans have agency against Latour's view that agency is deflated to networks and actants. Next is Fuller's treatment of Kuhn and Popper since Fuller views Kuhn as setting the stage for STS's ascendancy through STS's criticism of normativity in philosophy of science. Fuller's relation to STS to which Fuller is most closely associated is described. Fuller is critical of STS's lack of knowledge policy and of its relativism. STS's version of social constructivism is contrasted to Fuller's realism on the social sciences and social constructivism on the natural sciences.

Keywords Latour • STS • Kuhn • Popper

Science and technology studies (STS) is the name for a several kinds of studies of the social context of science different from traditional philosophy and history of science. Science and technology studies includes sociological and anthropological studies of science as well as studies of the literary form and rhetoric of scientific literature. Fuller is critical of recent STS's feigned neutrality and refusal to evaluate or criticize the results of science.

F.X. Remedios, V. Dusek, *Knowing Humanity in the Social World*,
https://doi.org/10.1057/978-1-137-37490-5_2

This chapter starts with Fuller's background in the history and philosophy of science. Next the debate between Fuller and Bruno Latour on the distinction or lack thereof between the humans and non-humans is recounted. Fuller's defense of the human with respect to epistemic agency is Latour's defense of actor-network theory in which agency is deflated into the network and attributed to invertebrates and machines as well as humans as "actants." Next there is a discussion of Fuller's normativity and STS, unity of knowledge, reduction and social participation in science, the claimed role of tacit knowledge and trust in scientific discovery, and the conflicting views of Thomas Kuhn and Karl Popper. Finally, the parallels drawn by Fuller between science debates and religious debates are described.

Fuller's background is different from that of most philosophers of science as well as that of constructivist sociologists of scientific knowledge, which may in part account for the great originality of his approach. Most philosophers of science and sociologists of scientific knowledge started out as science majors or scientists and, because of either impatience with drudgery in the lab or difficulty of the mathematics, combined with deepening fascination with the epistemological issues, shifted into philosophy. The more cynical would claim that many philosophers, historians, and sociologists of scientific knowledge are simply failed scientists. Fuller, in contrast, was a double major in history and sociology as a Columbia undergraduate and salutatorian of his class. He then studied history and philosophy of science at Cambridge and at Pittsburgh, at that time two of the best programs in the world in philosophy and history of science. Fuller's work is socially and historically richer by far than that of most philosophers and far more philosophically sophisticated than that of other sociologists (even those with philosophical training).

Fuller's breadth of reading in the humanities and social sciences is extraordinarily broad. He is knowledgeable about sociology, social history, anthropology, psychology, and a bit of economics, particularly development economics. He also has extensive familiarity with the literature of law, theology, rhetoric, history, philosophy, history of science, literary criticism, library science, political science, organizational behavior, and many other fields. In philosophy of science, he is familiar with all the schools of Anglo-American and continental European philosophy. Historically, his range is far wider than most philosophers of science, including medieval scholastic philosophy. He deploys his learning with wit and panache in an endlessly ironical dialectic. His breadth of knowledge of sociology, and social, economic, and general intellectual history, which he deploys while doing philosophy of science far more extensive than that of any philosopher with whom we are acquainted.

Imre Lakatos once said that there are two methods of the philosophy of science: "the method of logic and the method of gossip," both used by him. Fuller has mastered of both philosophical dialectic and "the higher gossip" with an uncanny ability to find the educational and social roots of thinkers: natural scientists, social scientists, and philosophers to shed light on the roots and motivations of their thought.

Fuller Versus Latour on the Uniqueness of Human Agency Versus Latour's Non-human "Actants"

Despite the existence of several schools of STS, the Paris School led by Bruno Latour is the main competitor of Fuller's social epistemology. Fuller has various criticisms of Latour, but a major one concerns Latour's eliding the distinction between humans and other natural entities. Latour treats not only humans but animals and machines as "actants." The agency of humans is dispersed and diluted in a network of humans, other organisms such as scallops, and physical artifacts, such as a doorstop. The network dissolves any special status to human agency that contrasts with various other forms of physical causality. Latour, in his later writings on physical ecology, speaks of "the parliament of things" in which all living things participate (Latour 1993, pp. 144–5; 2004b, p. 227). To the objection that non-human animals do not speak and represent themselves, Latour claims often humans act as "ventriloquists" for other humans and that humans speaking for non-human animals and plants are not problematical for his idea of a "parliament of things."

Fuller views human agency is different from the causality of other objects that Latour calls "actants." Fuller's treatment of this issue and his emphasis on the special nature of the human epistemic agent are central to the account of human nature behind his social theory of knowledge. In his more recent work, Fuller has moved from the sociology of science to an account of what humans must be to buttress his social epistemology. Fuller's turn from pure sociology of science to a kind of metaphysical, and even theological, approach is to buttress his account of humans as activist and creative and to an approach which is a modern, naturalist, version of the vision of humans as creating their knowledge and their world like that found in the early nineteenth-century German idealism of Johann Gottlieb Fichte and Georg Friedrich Hegel. Furthermore Latour's account of the humans entangled in an all-encompassing network is, notes Fuller, like that described in Hannah Arendt's portrayal of totalitarianisms in which responsibility is totally dispersed and diluted out of existence. Fuller does not mention the extreme cases of this total denial of responsibility by the Nazis, such as Albert Speer (technology did it) and Adolf Eichmann (the social network did it) (Arendt 1951, 1963).

Fuller does not deny Latour's claim that humans are part of the natural world or that they are enmeshed in it; Fuller allows that computers may become scientific agents and through strong artificial intelligence (AI) become human agents or that humans through biotechnology, nanotechnology, and AI may become embodied in physical forms quite different from our present biological constitution. However, Fuller notes that, though all the classical sociologists agree with the claim that humans are part of the natural world, they claimed that the social, in the sense of the human community, involves a special and unique sort of human activity (Barron 83). He wishes to build on this classical sociological tradition rather than on sociobiology or on a posthumanism that denies human agency.

Fuller's Normative Versus STS's Descriptive Theory of Scientific Knowledge

"My own project in 'social epistemology' has been dedicated to bridging this very significant gap between the 'is' and the 'ought'" (Fuller 2007a, p. 213). Fuller's approach to social epistemology combines two aspects that go beyond both traditional philosophy of science and the usual approaches to sociology of science. Fuller presents a transformed application of normative standards. Mainstream philosophers of science discuss norms in the logic or norms in the methods of science. However, they usually neglect both the normative as well even descriptive aspects of social organization of science and its relation to the larger society. Contemporary sociologists of scientific knowledge deploy a purely causal, non-evaluative approach to the social dynamics of science. Fuller advocates normative evaluation of the social impact and worth of scientific projects. Science and technology studies started to present itself as "radical," but it has become more and more detached and distanced from the normative issues involved in scientific controversies (Fuller 2005).

Part of the reason for this abandonment of the critical approach by mainstream science studies (well described by Fuller) is the financial and professional insecurity of younger proponents of science studies: "academically marginal contract based teachers…'the unfaculty'" (Fuller 2005, p. 76). He accounts for the social basis of popularity of one major program of production case studies in science and technology studies as "… one of the great make-work schemes for researchers when academia is imperiled" (Barron 2003, p. 87). The originators of science and technology studies, often members of such groups as the British Society for the Social Responsibility in Science and in the USA Science for the People, were mostly scientists and technologists who became disillusioned with the military and corporate uses to which their science was put and who wished to change the culture of science. These were what Fuller calls the "low church" of science and technology studies. As science studies became an academic specialty within social science and humanities, the "high church," departments, STS lost most of its critical and activist edge.

This detachment of STS from taking sides in scientific controversies to judge who is correct is not wholly implemented. STS feigns neutrality on controversies concerning fringe science, giving equal time or consideration to the challenging non-mainstream position and the mainstream one, as in considering debates concerning parapsychology, one does not find science studies being equally neutral on controversies such as that of claimed racial differences in IQ. In this STS shows that its apparent neutrality is sometimes a pose, belied by implicit political and cultural commitments heartily denied.

Latour's slogan is "follow the scientists" and he rejects the notions of critical evaluation and depth accounts of motivation. As Fuller advocates knowledge policy and knowledge policy analysts, his slogan for the role for knowledge managers might be "lead the scientists." This leading involves social evaluation of the allocation of resources to competing research programs. Fuller has written extensively on the topic knowledge management (Fuller 2002). At present, evaluation is done almost solely by leaders of scientific funding institutions, but Fuller wishes to extend this to further public involvement (Fuller 2002).

Fuller's advocacy of the critical approach to science advocated by Karl Popper (Fuller 2004) goes against the whole spirit of Latour's work as well as that of science studies. A further criticism of science studies as a field has been of its refusal to make evaluations not only of scientific theories but also of the social situation of science it describes.

Latour rejects, often contemptuously, hidden social forces and ideology (2004b). According to him nothing is hidden. There is no false consciousness, nothing beyond the actors' awareness of their situation. The social analyst has no deeper or truer insight into the situation than the participants. Latour opposes the critical approach (2004a) because it presupposes a hidden true realm on the basis of which common opinion and belief can be criticized. As part of this approach, Latour rejects Marxism and other sociological theories that claim that there are hidden social forces accounting for what is thought. Similarly, the whole intellectual movement of criticism or critique is dismissed as one of the negative developments of modernity. Latour's approach makes a virtue of this indifference to evaluating common belief and to attempting to look behind it for motives or social tendencies. However, in seeming contradiction to this, in his recent book, *Inquiry into the Modes of Existence: An Anthropology of the Moderns*, Latour frequently claims that the moderns are in a terrible state (Dusek 2014). This evaluation hardly fits with his neutrality with respect to science. Latour has come out as more and more conservative. He even, only half in jest, quotes Margaret Thatcher's claim society does not exist (2007, p. 5). In *Reassembling the Social*, he takes what would be called an extreme methodological individualism, though he would deny his units are individuals, but rather phases of processes.

Fuller is critical of the trendy turn of science studies people, such as found in Donna Haraway, Andrew Pickering, and Latour to the process philosophers A. N. Whitehead and H. Bergson, whom Fuller calls "funky" (2005a, b, p. 78) and whose work he sees as "metaphysical marshlands" (2007a, p. 212). Fuller grants that these philosophers may have made valuable contributions elsewhere, as in Bergson's notion of the "open society," as used by Karl Popper (2006a, p. 159; 2007a, pp. 46, 49) but sees postmodern STS's appeal to their metaphysics generally to be a retreat from genuine social criticism. Ironically science studies people such as Latour, Donna Haraway, and Andrew Pickering take Whitehead's process philosophy as digested by French writers and not from the original Anglo-American source. Latour's late turn from science studies to an obscure set of metaphysical claims, which has dismayed even some of his loyal followers, is an extreme case of this tendency (Latour 2013).

Fuller's Alliance with Logical Positivism

In contrast to Fuller's highly critical evaluation of mainstream science studies, he holds logical positivism in high regard. That Fuller admires the logical positivists and castigates present science studies may seem surprising.

His social and historical approach to the sciences strongly contrasts with most of the positivists' emphasis on pure formalism of mathematical logic. However, he aligns with the positivists on many issues. One thing that Fuller admires in the logical positivists is their normative approach to the formulation of actually existing science. They do not simply accept the scientists' often incomplete expressions of their science. The logical positivists wish to more rigorously reconstruct science and to foster science's drive toward unification. Ironically, Fuller was castigated by the so-called Science Warriors (scientists defending the social authority of science and opposing science studies) who often claimed to be defending a half-understood "positivist" view of science against the sociologists of scientific knowledge. The Science Warriors claimed the latter to be subjectivist and anti-science. In fact, in contrast to other science studies scholars, Fuller defends many features of the logical positivist normative account of science, such as the unity of science and the use of behaviorism in psychology. He also vigorously advocates the critical approach of Karl Popper against Thomas Kuhn's totally uncritical acceptance of contemporary science (Fuller 2004).

Fuller supports the logical positivists' defense of the traditional unity of science against more recent advocates of disunity of science, such as John Dupré and Peter Galison (2007a, pp. 11–52). Fuller also criticizes those philosophers of science who investigate the particular sciences, such as biology or quantum mechanics, when they describe different scientific methods for particular sciences, but act only as "underlaborers" (Fuller 2000a, pp. 260–261; 2006a, p. 48; 2011, p. 165). He calls them "philosophers *for* science" rather than philosophers *of* science (2004, p. 49; 2015, p. 145), that is, too often they promote present sciences rather than evaluating them or discussing their social management.

Kant, Neo-Kantians, and the (Dis)Unity of Knowledge

The neo-Kantians were late nineteenth-century philosophers in Germany who revived Immanuel Kant's theory of knowledge, in which the mind constructs and organizes sense experience. However, the neo-Kantians gave up much of Kant's ideal of the unity of knowledge. Fuller rightly and insightfully ties neo-Kantianism to its social base. Late nineteenth- and early twentieth-century neo-Kantians were academics in progressively specialized and compartmentalized universities. One might compare their projection of the organization of the differences among the academic disciplines in the structure of reality to the sociologist Émile Durkheim's claim that band that lived in circular or square villages envisioned the universe as circular or square, respectively.

This is separation of the disciplines reflected in the neo-Kantian Cassirer's otherwise very admirable book, *The Problem of Knowledge* (1944), to which Fuller refers several times (1988, p. 150, 202; 2007b, p. 23). The earlier volumes of the series (available only in German) (Cassirer 1906–7, 1910) treat philosophy and science from the fifteenth to early nineteenth century as a unity. The fourth volume, on the century from 1831 (death of Hegel) to 1932 (the rise of Hitler), presents philosophy in terms of philosophy *of* special disciplines: mathematics, physics, biology, and history, treated separately, even though each laid claim to an overall world view. Cassirer writes that by 1831 one could not treat all of knowledge as a unity (Cassirer 1951, p. 10, 15). Also, Cassirer leaves out the socially or personally relevant thinkers of the period, Marx, Kierkegaard, and Nietzsche, as well as the Kant-influenced philosophical sociologist Georg Simmel.

The neo-Kantians generally made a sharp distinction between the natural science and *geisteswissenschaft* (sciences of the spirit) or *kulturwissenschaft* (sciences of culture). This distinction was made in various ways. One was in terms of *verstehen* or intuitive understanding of meaning. Another was Wilhelm Windelband's focus on unique individuals (idiographic) versus focus on laws of nature (nomothetic). Yet another was Heinrich Rickert's focus on values as distinguishing the humanities from the natural sciences. The logical positivists denied these distinctions and were in favor of a unity of natural and social sciences.

However, the neo-Kantians' naive projection of the division of academic departments into the realm of knowledge was *not* shared by Kant. Clearly Kant distinguished psychology or anthropology from physics, but the overriding goal was the unity in knowledge. This is a regulative ideal above the specific constitution of knowledge of objects and causes. And behind this regulative ideal were normative considerations. The pursuit of unity of knowledge in some accounts is tied to a kind of ethical categorical imperative. Kant's Appendix to the Transcendental Dialectic can be construed as claiming all knowledge, including theoretical knowledge, is practical (Mudd 2013). Kant's romantic successor Gottlieb Fichte, whom Kant initially praised and then vigorously disowned on other grounds, emphasizes and makes much more explicit this priority of the practical.

This led to the titling of philosopher and social psychologist George Herbert Mead's late paper *The Philosophy of the Act*. Fuller's norm of unity of science, though mainly theoretically deriving from the logical positivists, and educationally from Goethe and Hegel, has a strong resemblance to Kant's normative orientation and tie to ethics. Fuller's social epistemology with an ethical background for the striving toward the unity of science continues these traditions.

Fuller also rejects, as did the logical positivists, four positions popular today:

1. The use of interpretive understanding in the sense of verstehen as a special method for the social science. Rather he supports a more behaviorist approach.
2. The appeal to tacit knowledge of Polanyi in philosophy of science as well as its use by strong opponents, such as Hubert Dreyfus, of strong artificial intelligence, the replication of conscious beings in silicon (Fuller 1996, p. 93, n. 7). The rejection of Dreyfus' critique supports Fuller's transhumanism where strong artificial intelligence will exist and human minds might be uploaded into computers.
3. The "disunificationists" denial of the unity of science. Fuller even claims that the notion of autonomous levels of organization (i.e. molecules, organisms, populations) and emergence is socially motivated by a desire to preserve academic specialties (Fuller 2007b, p. 134).
4. The role of "underlaborer" in contemporary philosophy of science, in which the philosopher simply tweaks the exposition and formulation of existing particular scientific theories, but does not evaluate or attempt, let alone attempt the reform, the scientific field as a whole.

Fuller is especially critical of the last position and role for the philosopher of science (2000b, pp. 260–263; 2006a, p. 45, 48). The term "underlaborer" was used by the philosopher John Locke, who as a theorist of knowledge claimed to be toiling under the theoretical physics of "the incomparable Mr. Newton" as merely "an under-labourer in clearing the ground a little and removing some of the rubbish that lies in the way to knowledge" (Locke 1894 [1689], p. 14).

The logical positivists' mentor Ernst Mach at the turn of the twentieth century produced "historical and critical" studies of sciences, particularly Newtonian mechanics. The followers of Ernst Mach, the logical positivists, did do some underlaborer work in formalizing theories in terms of mathematical logic, but their advocacy of total unity of science and often reduction of the biological and social sciences to something like physics made demands that went beyond the contemporary state of science. Karl Popper and his followers went even further, in the critical approach to science, criticizing aspects of probability theory, subatomic physics, and evolutionary biology. Popper's students Joseph Agassi and Paul Feyerabend went yet further in criticizing the state of science. However, Anglo-American philosophy of science since the 1980s turned mainly to cleaning up details of scientific theories such as quantum mechanics and evolution for the sake of the sciences and scientists, rather than evaluating, let alone criticizing, scientific theories, examining their social roots, and drawing out their social implications. This is evident in the last few decades of the leading American journal *Philosophy of Science* and various, more recent journals specializing in the philosophy of physics or the philosophy of biology.

Fuller's Critique of Kuhn

Part of Fuller's defense of a normative approach to the social structure and larger social environment of science as well as his criticism of the major source of the sociology of scientific knowledge is his scathing critique of the work of Thomas Kuhn.

It is well recognized what a tremendous influence Thomas Kuhn's *Structure of Scientific Revolutions (1970)* has had on social studies of science over the last half century. Kuhn was a major source of influence on the sociology of scientific knowledge (SSK). Besides Latour's Paris school, Kuhnian SSK is a main alternative program to that of Fuller in the sociology of science. Because of what he sees as a highly pernicious influence on science studies, Fuller launched an extensive, magisterial, and is intended to be devastating attack on Kuhn's account of science, *Thomas Kuhn: A Philosophical History for Our Times* (2000). In a detailed genetic account of Kuhn's setting, Fuller emphasizes Kuhn's early career and protection at Harvard was dependent on James Bryan Conant, who played an important role in the post facto justification of Hiroshima (Fuller 2000a, p. 174) and one of the major architects of post-WWII Cold War science. Conant initiated a course for non-science majors in which Kuhn was an instructor. Kuhn's participation and Conant's mentoring contributed to his shift to history of science and to *Structure* (Fuller 2000a, Ch. 4).

In a shorter, later follow-up to Fuller's account of Kuhn, *Popper* versus *Kuhn*, Fuller defends Popper's normative and critical approach against the supposedly descriptive and certainly uncritical approach of Kuhn. Fuller supports Popper's "insistence on an explicitly normative standpoint. ... though I am more explicit...about the political...means by which this normative standard is determined" (Fuller 2000a, p. 394, n. 27).

Fuller compares Kuhn's refusal to admit the association with and consequences of his work with Cold War attitudes and to Heidegger's decade long evasion of dealing with his Nazism (2004, pp. 110–122). Because of this over the top accusation of a strong guilt by association, and the shock at such strong criticism of their hero, some in science studies have claimed Fuller's demolition of Kuhn as simply outrageous.

However, even if hyperbolic, Fuller has a point here. Fuller traces in detail the most important novel development in the social structure of science, Big Science, of which the Manhattan Project was the paradigmatic exemplar. Kuhn himself seems not to have been happy with the direction of physics during and after WWII, but he never publicly discussed this. (Rather, in his writings, Kuhn dealt with the science of "great amateurs" to use Coolidge's term with respect to mathematics (Coolidge) from the 1600s to the early 1900s, but ceased his discussion of science with the WWI relativity theory and the quantum mechanics of the late 1920s.) This period ended the last major scientific advances that did not involve the subsequent gigantic apparatus and gigantic social organization of WWII and Cold War Big Science. Kuhn did not ever grapple with post-WWII Big Science, as did historian Derek de Solla Price, nuclear physicist Alvin Weinberg, and others at the time.

Kuhn's history of science is in many ways an aestheticization of past science as an escape from the morally tainted and more perplexing dilemmas of modern Big Science. Fuller notes the relationship to aesthetician Aby Warburg (Fuller 2000a, p. 215) and earlier figures in the history of art such as Heinrich Wölfflin (2000b, pp. 52–56). This somewhat parallels continental Marxist's turn to aesthetics (as in first generation of the Frankfurt School) after the failure Communist mass movements (Anderson 1976).

Kuhn presents actually existing science as the most rational of human endeavors but avoids discussing the less scientifically rational political and economic motives that have driven post-WWII science. Kuhn makes very few remarks on social factors in the development of science. These concern Copernicus' relation to calendar calculations (1957, pp. 125–6;

1970, p. x, 69) and a single sentence and footnote in an article on class relations in the conflict of apothecaries and physicians in early modern medicine (Kuhn 1977, p. 159, n. 21). Kuhn's work lacked and distained to pursue work of others based on his dealing with relations of science and society. He was famously shocked and disturbed by criticisms of him by Popperians such as Lakatos and totally denied any association with the political and culturally radical students who made use of his work. In a further article on Kuhn, Fuller compares Kuhn to Chance the gardener in the novel and film *Being There* who makes inane statements taken by hearers to be profound truths (Fuller 1992). Chance is completely unaware of the tremendous effect of his utterances on others. This is obviously very unflattering, but it is true that Kuhn, despite his success, had no expectation or awareness of most of the effects of his book. He was taken aback by the use of his book by Marxists and by members of counterculture who rejected science, as well as those who applied his terminology to the arts and humanities, and by the sociologists of scientific knowledge. After the attack by Imre Lakatos, Kuhn retreated for a time to interviewing the founders of quantum theory (Kuhn et al. 1967). He later wrote a book on the beginning of quantum physics (1987) that is a purely descriptive history totally lacking any ideas from *Structure*. In his 1970 postscript to *Structure of Scientific Revolutions*, reacting to his critics, Kuhn moves, at least temporarily, to a position like that of philosophical foundationalism, inconsistent with the theory-laden treatment of evidence in the work to which it is added. Kuhn also grapples with linguistic philosophers who rejected linguistic relativism (Kuhn et al. 2000, pp. 35–38, 196–202). These philosophers emphasize reference rather than meaning. In doing so, Kuhn apparently jettisons Benjamin Lee Whorf's linguistic relativism that he cited in the Preface of *Structure* (Kuhn 1970, p. vi). There is a shift to emphasis on metaphor and analogy (Kuhn 1992, pp. 30–31, 196–207). Also, Kuhn drops the use of psychology of the gestalt switch in understanding paradigm change. He also drops his original but controversial and much criticized notion of incommensurability. Originally this was the claimed inability to compare alternative paradigms or theories, because of not only different meanings of the apparently on the face of them are the same words or symbols but because of different standards of evaluation and different scientific methodologies. He replaces this with a simple change of a lexicon of terms classifications of entities. The richer aspects of methodological and evaluative incompatibility, present in *Structure*, are dropped. In Kuhn's last book-length work, on Max Planck (Kuhn 1987), Kuhn made no use of the theory he presented in *Structure*.

In Kuhn's Rothschild lecture that the Science Warriors (opponents of social and literary studies of science) passed around among one another like a secret, holy manuscript, Kuhn strongly disassociated himself from postmodernism of many of his humanistic and social science followers. He calls the work of David Bloor and the Edinburgh School of sociology of science "postmodernism gone mad" (1992, p. 7; 2000, p. 110). In all, the later Kuhn retreated from the provocative and very controversial theses of his original work.

One notion that Kuhn appeals to is Michael Polanyi's tacit knowledge. The tacit nature of this knowledge prevents an articulation and analysis, and hence prevents criticism, of the overarching paradigms that Kuhn claims guide normal science.

Fuller's Behaviorism and the Rejection of Tacit Knowledge

Fuller is sympathetic to behaviorism and rejects verstehen and hermeneutics. The latter two approaches use intuitive and empathetic understanding of other persons and their literary products, respectively. Fuller, who refers to "atavism of phenomenology" (Fuller 1993, p. 47), appeals to psychological studies of self-knowledge to raise doubts about the capacity of the phenomenologist to gain objective and universal knowledge based on her intuitions.

Fuller distrusts tacit knowledge and considers appeals to it to be unfalsifiable and circular. Part of Fuller's rejection of tacit knowledge may be in part because of his opposition to the authoritarian and dogmatic conception of science in Polanyi's *Personal Knowledge* (1958), which Fuller considers highly ideological and Polanyi's earlier accounts of the social structure of science. Polanyi, rather like the romantic intuitionist Wilhelm Schelling, claims that the scientists' knowledge is unaccountable. Further, for Polanyi, like Schelling, the intuitions of the prestigious scientists must be taken on faith and their deliverances adhered to. Again, this is similar to the romantic cult of genius. Nothing could be further from Fuller's Popperian conception of science as an open and democratic enterprise, Fuller going further that Popper in emphasizing public accountability of science and public input to the value of its research projects. Appeals to expertise via tacit knowledge also appear in works of the sociologist of science of the leader of the "Bath School" of sociology of science, Harry Collins (2007, 2010).

Fuller's Admiration for and Use of Popper

Fuller strongly supports Popper's critical approach toward science as opposed to Kuhn's uncritical approach. This is most evident in his follow-up to his Kuhn book, *Kuhn* vs. *Popper: The Struggle for the Soul of Science* (2004). While Popper is today often portrayed at worst as an old positivist fuddy-duddy, or, at best, as an ahistorical formalist by most science studies people, Fuller grasps the insight and originality of Popper, even if he does not follow either Popper's attacks on the sociology of knowledge, or Popper's strict methodological individualism. Fuller notes, following to Hacohen's biography (2000; Fuller n.d., 2004, pp. 219–220), that the early Popper in Vienna was far to the left of the Popper of his later years. What Fuller takes from Popper, and his closest followers, such as Joseph Agassi, is a willingness to evaluate and criticize the social structure of science as well as the content of scientific theories. The Popperians are not Lockean underlaborers who always defer to the judgments of influential scientists even on the veracity of particular scientific theories or the infallibility of the social judgments of the scientific community. Popper and his followers are well aware the ideal of publicity of data, openness to criticism, and skepticism is often followed more in the breach than in the observance in actual science. Sociologist Robert King Merton had treated these professed ideals of scientists as actually being followed (Fuller 1997, pp. 62–63). Fuller compares this to sociologists always accepting the word of politicians or religious leaders that their professed values are put into practice. One aspect of Popper that Fuller probably does not accept is the critique of sociology of knowledge found in *The Poverty of Philosophy*, a book that Hilary Putnam in his radical stage once called the worst book ever written, apparently forgetting *Mein Kampf* and a few others (Putnam 1970). Popper's critique of "historicism" misrepresents the actual German historicists (such as Dilthey and Meinecke) as holding to laws of history of the sort that Comte and the more schematic expositions by Marx present, although Fuller would agree with Popper about the latter sort of deterministic historicism.

What Fuller takes from Popper is his advocacy of democracy in science and his dislike of the dogmatic and prestige-oriented actual scientific community that was praised as ideal by Polanyi.

Trust and Tacit Knowledge

Fuller's social epistemology rejects a strong strain of sociology of science and of expertise that appeals to "trust" and to the "tacit knowledge" of practitioners.

Michael Polanyi, the physical chemist turned social scientist turned amateur philosopher, was stimulated to develop his views by his encounter (Polanyi 1966, p. 3) with the Soviet official and theorist Nikolai Bukharin, whose visit to London led to the radicalization of a number of leading biologists and their attempts to develop a Marxist history and sociology of science, advocated the planning of science (Dusek 1990). To counter this Polanyi appealed to a view of science in which the intuitions of leading scientists, intuitions that cannot be analyzed or externally evaluated, are the guiding factors in the growth of scientific knowledge. Polanyi's notion of "Genius in Science" and of the hierarchy of authority in the scientific community is a twentieth-century version of Schelling's romantic conception of genius and intuition. The leading scientists, with their ineffable intuition, guide the lesser beings within the scientific community. Obedience to the authority of their intuitions and authority is central to the structure of science.

Thomas Kuhn explicitly borrowed Polanyi's notion of tacit knowledge from *Personal Knowledge* in *Structure*. The most extreme features of Polanyi's notion of authority in science appear in Kuhn's earlier article "The Function of Dogma in Science" (Kuhn 1957).

Polanyi's notion of science resembles the medieval belief in the divine right of kings or the absolute authority of church leaders. Fuller claims Kuhn and Polanyi held a "monkish" view of the scientist (Fuller 2004, p. 102).

Fuller points appeal to trust is a twenty-first-century version of nineteenth-century ideas of "habit" and "tradition." In conservative political theory, the bare fact of tradition and age is sufficient to legitimate norms and institutions. Edmund Burke and in the eighteenth century as well as late nineteenth-century British Hegelians appealed to sheer tradition and habit against the attempts by utopians, utilitarian reformers, and

French revolutionaries to rational re-organize society. This traditional conservative appeal to habit and custom as well as to sheer longevity has become less popular in political theory over the last century.

Fuller's behaviorist leanings and sympathy for experimental philosophy lead him to demand evidence in behavior for the claims to authority of ineffable intuitions and blind trust in these intuitions of the leading scientists. Examination of how these intuitions and accepted traditions change through time may reveal their real basis. Often justifications in terms of what used to be called habit, tradition, and custom, may in fact be cases of what Eric Hobsbawm and Ranger (1984) called "the invention of tradition," wherein innovations are justified by claiming them as deliverances of ancient (fictional) traditions.

Fuller calls appeal to trust in unarticulated deliverances a kind of phlogiston in the sociology and philosophy of science. Like the later rejected chemical component phlogiston is an unverifiable substance appealed to in explaining numerous phenomena and processes and appropriately modified in description to account for whatever is desired to be explained. Trust, like phlogiston's existence, can only be explained by its absence (Fuller 2006b, p. 173). He notes that "trust" and "tacit knowledge" have replaced the older conservative standbys of tradition, custom, and habit (Fuller 2006c, p. 45).

Fuller's sympathies with respect to the evaluation of intuitions are with what is now called experimental philosophy. This approach was attempted early by Arne Naess, but was neglected until recently (Naess 1938). In it intuitions are surveyed among the populace, rather than simply appealed to by the individual philosopher claiming the intuition (Sytsma and Buckwalter).

One of the positions of Fuller that goes against the grain of science and technology studies with its emphasis on tacit knowledge and trust is his sympathy toward behaviorism. "When push comes to shove, I rather like behaviorism…" (Remedios, p. 117). In the debate between the behaviorist B.F. Skinner and the linguist Noam Chomsky, Fuller says that Skinner's research program attempts "to derive an organism's behavior as a function of environmental change," while "Chomsky's work fails to meet this basic principle" (Barron 2003, p. 133). Fuller's behaviorism also applies to the scientist. His "thin" notion of the scientist makes the scientists qua scientist a purely calculating entity (Remedios, pp. 116–117). His "thin" version of the scientist is similar to microeconomics' model of "economic man." Fuller holds to Herbert Simon's satisficing model of decision theory seeking achievable satisfactory results, rather than the maximizing model of standard microeconomics. Nevertheless, Simon's model abstracts from other personal, emotional, and ideological aspects of the scientist that resembles to standard "economic man."

One irony of this is that Fuller, who is so concerned with recognition of the social embeddedness of science, treats the scientist as a pure methodological machine. This allows him to easily include computers as epistemic agents. However, it seems in conflict with Fuller's emphasis in the historical lineages he traces of scientists, constantly uncovering surprising connections with influential views that do not fit the standard stereotypes of scientists as purely rational hypothesis testers. It also implicitly (if not explicitly in Fuller) leads to a sharp distinction between "internal" and "external" factors in the growth of science, in contrast to the accounts of many sociologists of scientific knowledge, such as Bruno Latour, Steven Shapin, and, indeed, Fuller himself.

One could ask what sort of behaviorism is it with which Fuller sympathizes. With Fuller's vast knowledge of psychological theories and experiments, one should like to hear whether Fuller's "behaviorism" is like that of B. F. Skinner or that of Edward Chace Tolman, who was the model for his teacher Herbert Simon's approach, and what relation it has to the less scientific forms of philosophical behaviorism such as Gilbert Ryle's logical behaviorism or Ludwig Wittgenstein's behavioristic tendencies (the latter despite the fact that Wittgenstein used to pound the table in seminars when denying he was a behaviorist).

Fuller rejects on behaviorist grounds the notions of trust, expertise, and tacit knowledge, which he calls "social phlogiston." It depends on an unverifiable and unarticulable quality of the experts who have been trusted. It can be claimed to be present without any evidence. Denial of its presence can be morally criticized. Fuller notes that the "trusters" resemble those traditional conservatives who a century ago appealed to tradition and habit. The assumption was that if something was old, it was good. Also it is assumed that the normative orientation of society is correct. Similarly, with "the trusters," Fuller assumes that the normative order of science is correct, even if the norms cannot themselves be characterized.

Parallels of Scientific with Religious Disputes

Fuller is knowledgeable, not only of contemporary and early modern philosophy but of medieval philosophy in his work on philosophy and sociology of science. Except a few Catholic philosophers of science (Patrick Heelen, William Wallace, and Bas van Fraassen), the relevance of medieval thought and concepts to contemporary science is often dismissed (as in the uninformed Enlightenment conception of "the dark ages"). Fuller is unique in relating issues of modern science to disputes between Dominicans and Franciscans: Thomists and Scotists (see Chap. 4).

> Whatever one makes of these theological niceties today, the issues at play are not so very different from the controversies that have periodically flared up, both within science and between scientists and the historians, sociologists, and philosophers who study them. (Fuller 2010, p. 46)

Fuller notes the similarities of disputes both within science and between those in science and technology studies people and their scientist critics with religious controversies. For instance, when science studies people analyze the history, politics, and rhetoric involved in science controversies, the science studies researchers are described as anti-science. This resembles the reaction of religious believers to the so-called Higher Criticism of the Bible in the nineteenth century. Literalist Bible believers accused those who undertook historical and linguistic analyses of the Bible of atheism. Most of the higher critics of the Bible were themselves religious. Most of the leading science studies people have scientific backgrounds and have high regard for science (Fuller 2006a, p. 1190). To study science as a social institution is not to be anti-scientific. In another parallel between science studies and religious differences, Fuller distinguishes within science and technology studies between the "low church" and "high church" tendencies (Fuller 2000b, pp. 409–412; Barron 2003, p. xii). In Anglicanism the low church deemphasized priestly authority and emphasized evangelism, while the high church emphasized priestly authority and ritual. The "low church," characteristic of the earlier days of science studies, agreed with those who criticized and attempted to change the state of science and technology. Groups of scientists, such as Science for the People members, lent their expertise, for instance, in campaigning against science's subservience to the military. On the other hand, the "high church" approach corresponds to the present academic, value-neutral, and postmodern aspects of the more recent science studies. Fuller opposes this shift to a stance that drops any evaluation of scientific research and technological projects.

Another parallel between the social situation of science and that of religion is what Fuller calls the development in recent decades of "protscience" on analogy to Protestantism. Protestantism gave the lay people direct access to the Bible, which had been restricted primarily to priests in Roman Catholicism. Similarly, contemporary "protscience" involves the public being able to know and sometimes participate in and criticize vetting scientific and technological projects. It also involves the public's embracing of movements that many orthodox scientists claim pseudoscience, such as acupuncture and holistic medicine (Fuller 2010, pp. 61–71, 64–67).

In the theory of belief, Fuller contrasts the Franciscan "believe in order to understand," which he associates with Kuhn, and the acceptance of a claim as a decision, as in Popper (2004, pp. 64–67). Fuller notes that Popper early was a reader of Kierkegaard, who emphasized leaps of commitment (2004, pp. 62–63).

Fuller also writes that many theological notions, such as priesthood (scientists and equivalents of priests from Comte to nuclear advisors), eschatology (future utopia), soteriology (science as salvation), and theodicy (justification of harm in terms of overall good system), are implicitly present in science today but are not called such or even recognized as such by scientists and the public (Fuller 1997, pp. 48–62).

Fuller's social epistemology approach to the study of science and technology is highly original. It incorporates several positions and approaches that are usually thought to be antagonistic to one another. Centering on epistemic agency, it emphasizes the active and evaluative role of the knower. This combines aspects of the German idealist tradition with elements of its apparent arch enemy Karl Popper. He defends the positivists' quest for unity and reduction and combines Popper's critical approach to science with the explanation of scientific theories and technological projects in terms social and historical context.

References

Anderson, P. (1976). *Considerations on Western Marxism.* London: Verso Books.

Arendt, H. (1951). *The Origins of Totalitarianism.* New York: Harcourt Brace.

Arendt, H. (1963). *Eichmann in Jerusalem: A Study in the Banality of Evil.* New York: Viking Press.

Barron, C. (2003). A Strong Distinction Between Humans and Non-Humans Is No Longer Required for Research Purposes: A Debate Between Bruno Latour and Steve Fuller. *History of the Human Sciences, 16*(2), 77–99.

Cassirer, E. (1906–1907, 1910). *Das Erkenntnisproblem in der Philosophie und Wissenschaft der Neueren Zeit* (Vols. 1–3). Berlin: Bruno Cassirer Verlag.

Cassirer, E. (1944). *The Problem of Knowledge.* New Haven: Yale University Press.

Cassirer, E. (1951). *The Philosophy of the Enlightenment.* Princeton: Princeton University Press.
Collins, H. (2007). *Rethinking Expertise.* Chicago: University of Chicago Press.
Collins, H. (2010). *Tacit and Explicit Knowledge.* Chicago: University of Chicago Press.
Dusek, V. (1990). The Bukharin Delegation on Science and Society: Action and Reaction in British Science Studies. In N. N. Kozlov & E. D. Weiss (Eds.), *Nikolai Ivanovich Bukharin: A Centenary Appreciation* (pp. 129–147). New York: Praeger Publishers.
Dusek, V. (2014). [Review of] Bruno Latour, *An Inquiry into the Modes of Existence. Notre Dame Philosophical Review.* http://ndpr.nd.edu/news/an-inquiry-into-modes-of-existence-an-anthropology-of-the-moderns/
Fuller, S. (1988). *Social Epistemology.* Bloomington: Indiana University Press.
Fuller, S. (1992). Being There with Thomas Kuhn: A Parable for Postmodern Times. *History and Theory, 31*(3), 241–275.
Fuller, S. (1993). Social Constructivism Teaching Itself a Lesson: Science Studies as a Social Movement. *Danish Yearbook of Philosophy, 28*, 47–60.
Fuller, S. (1996). Social Epistemology and Psychology. In W. O'Donohue & R. Kitchener (Eds.), *The Philosophy of Psychology.* London: Sage Publications.
Fuller, S. (1997). Making Science an Experimenting Society. In W. Dunn (Ed.), *The Experimenting Society: Essays in Honor of Donald T. Campbell* (pp. 69–102). New Brunswick: Transaction Publishers.
Fuller, S. (2000a, February 22). E-mail to Remedios.
Fuller, S. (2000b). *Thomas Kuhn: A Philosophical History of Our Times.* Chicago: Chicago University Press.
Fuller, S. (2002). *Knowledge Management Foundations.* Boston: Butterworth-Heinemann.ß.
Fuller, S. (2004). *Kuhn vs. Popper: The Struggle for the Soul of Science.* New York: Columbia University Press.
Fuller, S. (2005). Is Science Studies Revolutionary or Merely Revolting? *Science Studies, 18*(1), 75–83.
Fuller, S. (2006a). Humanity as the Site of Ideological Conflict in the Twenty-First Century. *Ludus Vitalis, 14*(26), 65–88.
Fuller, S. (2006b). *The New Sociological Imagination.* London: Sage Publications, Ltd.
Fuller, S. (2006c). *The Philosophy of Science and Technology Studies.* New York: Routledge.
Fuller, S. (2007a). *New Frontiers in Science and Technology Studies.* New York: Polity Press.
Fuller, S. (2007b). *The Knowledge Book: Key Concepts in Philosophy, Science, and Culture.* Montreal: McGill-Queens University Press.
Fuller, S. (2010). *Science.* Durham: Acumen Publishing Limited.

Fuller, S. (2011). *Humanity 2.0: What It Means to Be Human Past, Present and Future*. London: Palgrave Macmillan.
Fuller, S. (2015). Retrieved January 28, 2017 from https://twitter.com/professorstevefuller/status/660426459547242497
Fuller, S. (n.d.). Steve Fuller on Malachi Cohen. http://www.the-rathouse.com/shortreviews/Fuller-on-Hacohen.html
Hacohen, M. H. (2000). *Karl Popper: The Formative Years*. Cambridge: Cambridge University Press.
Hobsbawm, E., & Ranger, T. (1984). *The Invention of Tradition*. Cambridge: Cambridge University Press.
Kuhn, T. (1977). *The Essential Tension*. Chicago: University of Chicago Press.
Kuhn, T. (1992). *The Trouble with the Historical Philosophy of Science*. Robert and Maurine Rothschild Distinguished Lecture, 19 November 1991. An Occasional Publication of the Department of the History of Science. Cambridge, MA: Harvard University Press.
Kuhn, T. S. (1957). The Function of Dogma in Science. In M. Claggett (Ed.), *Critical Problems in the History of Science* (pp. 347–369). Madison: University of Wisconsin Press.
Kuhn, T. S. (1970). *The Structure of Scientific Revolutions* (2nd ed.). Chicago: University of Chicago Press.
Kuhn, T. S. (1987). *Black Body Theory and the Quantum Discontinuity*. Chicago: The University Press.
Kuhn, T., et al. (2000). *The Road Since Structure*. Chicago: The University of Chicago Press.
Kuhn, T. S., et al. (1967). *Sources for History of Quantum Physics*. Philadelphia: American Philosophical Society.
Latour, B. (1993). *We Have Never Been Modern*. Cambridge, MA: Harvard University Press.
Latour, B. (2004a). Why Has Critique Run Out of Steam? *Critical Inquiry, 30*, 225–248.
Latour, B. (2004b). *The Politics of Nature*. Cambridge, MA: Harvard University Press.
Latour, B. (2007). *Reassembling the Social: An Introduction to Actor-Network-Theory*. Oxford: Oxford University Press.
Latour, B. (2013). *An Inquiry into Modes of Existence: An Anthropology of the Moderns* (trans. Porter, C.). Cambridge MA: Harvard University Press.
Locke, J. (1894). *An Essay on Human Understanding*. Oxford: Clarendon Press. (Orig. 1689).
Mudd, S. (2013). Rethinking the Priority of Practical Reason. *European Journal of Philosophy, 23*, 1–25.

Naess, A. (1938). *'Truth' as Conceived by Those Who Are Not Professional Philosophers*. Oslo: I kommisjon hos J. Dybwad. Skrifter utg. av det Norske videnskaps-akademii Oslo, II. Hist.-filos. Klasse.
Polanyi, M. (1958). *Personal Knowledge: Towards a Postcritical Philosophy*. Chicago: Chicago University Press.
Polanyi, M. (1966). *The Tacit Dimension*. Garden City: Anchor Books.
Putnam, H. (1970, December 31). Personal Communication with Dusek.

CHAPTER 3

Fuller's Social Epistemology and Epistemic Agency

Abstract This chapter is on agent-oriented social epistemology, which emphasizes epistemic agency or the knower as ontologically open. This is from Fuller's move to transhumanist in which to knower is enhanced to become disembodied. Fuller views the epistemic agent to make knowledge to act in the world as contrasted to analytic social epistemology's epistemic agent, who is a human knower with beliefs and does not make knowledge through construction of reality. There is also a discussion of cognitive economics in which the epistemic agent makes knowledge and leverages beliefs to action instead of the epistemic agent having beliefs to access knowledge.

Keywords Transhumanism • Agency • Agent-oriented social epistemology

The previous chapter notes the difference between Fuller's notion of agency and Latour's deflationary actor-network theory where there epistemic agency is distributed to the networks and actants, which include humans, machines, and invertebrates. Epistemic agency is important to Fuller, who moves from an emphasis on a knowledge policy of distribution of knowledge to the question on what kind of being should the knower be? This chapter explores Fuller's epistemic agency in which the agent

F.X. Remedios, V. Dusek, *Knowing Humanity in the Social World*,
https://doi.org/10.1057/978-1-137-37490-5_3

makes knowledge to act in the world compared to analytic social epistemology's epistemic agent or knower as represented by Alvin Goldman, in which the agent does not make knowledge to act in the world. There are also discussions of the differences between agent-oriented social epistemology versus object-oriented social epistemology in which the agent-oriented social epistemology agent makes knowledge and in object-oriented social epistemology, the agent does not make knowledge. This chapter proceeds with a discussion of Fuller's version of social epistemology as a version of cognitive economics, as it is the epistemic agent who makes knowledge and leverages beliefs into action. This chapter also discusses expertise to contrast agent-oriented social epistemology, which does not defer uncritically to experts, and object-oriented social epistemology, which does so with respect to domains of knowledge. Then this chapter concludes with a discussion of Fuller's conception of thick and thin conceptions of agency.

Fuller's Social Constructivism and Analytic Epistemology

STS holds to a social constructivism in which science is socially constructed. Fuller accepts that science is socially constructed and asks the question how science is to be legitimated.

From 2000, Fuller has emphasized epistemic agency with social constructivism in which the agent makes knowledge to act in the world. This notion of maker's knowledge follows Bacon, who holds an intimate relationship between the objects of cognition and the objects of construction (Lynch 2001, 2016; Pérez-Ramos 1988), is different from the analytic social epistemology in which the agent is not active to make knowledge but rather goes from an inside out position in which the agent's belief is about epistemic objects, which represent the world. Epistemic success is for the epistemic agent to construct reality in a form of instrumental rationality.[1] Even with Goldman's analytic epistemology's casual theory of knowledge, the agent does not make knowledge but has a reliable connection of belief to truth. Fuller considers epistemology to be a style of metaphysics in which the epistemic agent, the knower, makes knowledge to act in the world as opposed to reality being fixed. On epistemology as a style of metaphysics, Fuller follows Hegel, whose phenomenology of mind is an epistemology which transcends subject and object to an idealist metaphysic, that is, the world is not portrayed as totally mind dependent but as in part constituted by the mind. Hegel presents knowledge in a series of stages, passing from bare sense certainty to reason and laws, culminating in what Hegel calls absolute knowledge, in which the knower and the known become one. Counter to Hegel and Fuller is the classical realist's (Goldman's) correspondence theory of truth, in which the epistemic agent does not make knowledge, but the epistemic agent gains knowledge through getting the objects of knowledge correct which represent the world. Fuller with his constructivist reading of Hegel, the epistemic agent makes knowledge (Remedios and Dusek 2016).

Fuller's Humanity 2.0, Changes and Expansions of Personhood and Agency

With humanity 2.0, Fuller expands personhood and views that the ontology of person to be open so that a "person" can refer to a human being as well as non-human entities, such as corporations, universities, and nation-states, which are disembodied. Common to all of them is that they are all producers of knowledge.

Fuller's transhumanism is the view that humanity can be enhanced or redesigned through technology. According to Fuller, converging technologies, such as biotechnology, nanotechnology, and computer technology, are transforming and enhancing humanity to humanity 2.0. The converging technologies agenda was triggered by 2002 National Science Foundation report to project a post-Cold War science policy agenda to get the above sciences working together to improve humanity to reduce medical and pension costs (Roco and Bainbridge 2003). With the impact of the technosciences such as AI and synthetic biology, the ontology of the agent is changing, and the knowledge they are making is changing. With transhumanism, the epistemic agent is enhanced and so may be transformed to be disembodied. The transhumanist knower will have exceeded the human boundaries of analytic social philosophy's knower who is more Cartesian and does not make knowledge. The maker's knowledge in which the agent intervenes in the world to make knowledge is different from theoretical knowledge, which is the concern of philosophers of science, or propositional knowledge, which is the domain of Goldman and analytic epistemology. Fuller differs with Latour and STS, which has a deflationary actor-network theory in which epistemic agent is distributed to the parliament of things and humans. Fuller would say that knowledge from the parliament of things is not scientific knowledge because the agent does not make knowledge through the construction of reality. Though Schmidt (2011) has argued that STS's representatives such as Haraway and Latour follow Bacon's maker's knowledge to make knowledge to act in the world, we argue that neither Haraway nor Latour as STS representatives focuses on epistemic agency or the knower to make knowledge through the construction of reality. Rouse (1996, 2002) argues that knowledge is deflated to scientific practices, which is foundational to agency. In response to Remedios (2003a, b), who compares Fuller to Rouse on the legitimation of science and normativity, Rouse (2003) argues that Fuller's and Remedios's view requires a notion of causation. Fuller's view is that of the maker's knowledge, in which the epistemic agent is active in making knowledge to construct reality, is that notion of causation is embedded in the agent who makes knowledge.[2]

Fuller's notion of person is based in the legal definition of "liability" in which a "person" (as either an individual or corporation) is held responsible for her or his actions (Fuller 2016a). STS has transferred responsibility from corporations, which are legal persons, to networks where responsibility is distributed across the network. For example, in actor-network theory, humans and non-humans as part of the network have the capacity to act (Latour 1987; Callon 1986).

For Fuller, epistemic agency, which includes humans and non-human entities such as corporations, nation-states, and universities, is a form of personhood. Though the epistemic agent is socially constructed, the standard by which the epistemic agent is evaluated is truth-oriented. For Fuller, truth is a systematic representation of reality, a grand unified theory of everything. To achieve this type of scientific knowledge, it is an open question as to the type of agent that would be most appropriate. It would not necessarily be an individual scientist such as Einstein or Hawking—whom some may consider to be a cyborg. A person may also be a computer or a group than humans and can literally end humankind. For Fuller, humanity would continue if homo sapiens end. Fuller's agent has to be legally recognized by others to produce knowledge. Epistemic privilege is granted. Goldman follows the Cartesian model that the agent recognizes herself as an agent.

Agent-Oriented Versus Object-Oriented Social Epistemologies

Fuller's view is that the individual versus the collective is not the most important issue in social epistemology, but the most important is agent versus object-oriented social epistemology. Fuller distinguishes between agent-oriented social epistemology in which the epistemic agent makes knowledge and object-oriented social epistemology in which the agent does not make knowledge and the nature of knowledge is divided and based on its objects. For agent-oriented social epistemology, the knower unifies knowledge based on a personal world view, like a medieval master of liberal arts or the polymath Goethe. Fuller is sympathetic to German idealism, in which the knowledge process is incorporated into the agent so that the agent's highest epistemic state lies in the achievement of the unification of everything in her or his mind in the manner of Goethe and Hegel. The question is: does the epistemic agent classify objects of knowledge according to knowledge domains, which is a neo-Kantian phenomenology of academic disciplines such as the natural sciences and the social sciences, or does the epistemic agent organize the inventory of objects of knowledge to achieve her or his ends of knowledge? Fuller's view is that the agent unifies knowledge by incorporation of the objects of inquiry into herself or himself through organization of the inventory and then provides a set of norms based on those objects upon which the agent can act.

For object-oriented social epistemology, the unity of knowledge, which is based on social integration of different forms of knowledge, is achieved via the trading of knowledge by experts of discrete domains. Fuller's agent-oriented social epistemologist develops knowledge policy according to the governance of science. Fuller favors agent-oriented social epistemology over object-oriented social epistemology, because social epistemology concerns how knowledge can be pursued normatively to act in the world (Fuller 2014, pp. 14–15). Analytic social epistemology and STS tend more towards object-oriented social epistemology in which the epistemic agent relies on expertise of others.

As agent and object-oriented social epistemologies are ideal types, the distinction between the two ideal types may be exaggerated to highlight the differences between them concerning the normative implications in the history of philosophy. For example, agent-oriented social epistemology is characterized as idealist/rationalist, and object-oriented social epistemology is characterized as empiricist, in that agent-oriented social epistemology is based on the agent's mind constructing knowledge and object-oriented social epistemology is based on experimenting to obtain knowledge. Agent and object-oriented social epistemology can be considered to be two ends of a continuum in the same way that individual and collective epistemologies can be considered to be the opposite ends of a continuum (Remedios and Dusek 2016).

Cognitive Economics

Fuller views that social epistemology is a version of cognitive economics, which is an economics of knowledge following Rescher (1989) and Peirce (1958). Cognitive economics is part of Fuller's social epistemology, which is knowledge policy on how knowledge should be organized. Cognitive economics is about the economics of knowledge made by the agent for construction of reality. The epistemic agent, who is pulled in different directions due to social forces, makes her or his decisions on actions to take in response to the social forces and the consequences of those actions.

Fuller notes that there are two kinds of cognitive economics of social epistemology, demand-side and supply-side. This terminology comes from economics. Demand-side economics focuses on consumption and the encouragement of consumption to boost the economy. Supply-side economics emphasizes the investment and production side of economics rather than consumer demand. The terminology of "supply-side" policies came during the US administration of Ronald Reagan. It was used politically as a justification for cutting taxes on the rich in order to aid the poor by "trickle down" economics. This was counterposed to Keynesian economics which focuses on effective demand and government investment. Demand-side social epistemology concerns the demand for knowledge based on the need for knowledge, and supply-side social epistemology concerns the supply of knowledge based on evidence for knowledge.

Fuller contrasts demand-side and supply-side epistemologies. Demand-side epistemology is characterized by James' "Will to Believe," in which the epistemic agent gains knowledge and voluntarily believes to leverage belief into action to construct reality. Supply-side epistemology is characterized by W. K. Clifford for whom ethical knowers place their knowledge on a foundation to protect them from an uncertain future (Clifford, pp. 176–7). Fuller, who is a demand-side epistemologist, holds that the epistemic agent decides on a hypothesis to be tested. A demand-side future, which is proactionary, is one of growth and flourishing. A supply-side future, which is precautionary, is one of avoiding risks.

Fuller's example of the precautionary side is Elgin's notion of an epistemic agent who relies on experts to set authority on a domain of knowledge and then believes the conclusions of the expert unless the agent has independent knowledge (Elgin 2013; Fuller 2016b). Elgin's notion of belief is suspect because, based on a different interpretation of evidence, the agent can reject the conclusion of the expert and because having a belief does not imply that the agent should act on it. Fuller's view is that the agent should take on information and then decide whether to believe that information and to act on it.

Fuller's preferred epistemic agent follows proactive rather than reactive or precautionary principles. This leads her or him to advocate greater risks on the part of scientific researchers and the encouragement of arrangements whereby people can knowingly volunteer as subjects of risky or dangerous experiments. He rightly claims the understandable reaction to Nazi experiments on concentration camp prisoners and the American Tuskegee experiment of letting syphilitic African Americans die when cures for the disease became available which led to extremely circumscribed boundaries on experiments on human subjects. Additional examples that Fuller mentions in a review of "Critical Neuroscience" (Fuller 2013) are José Delgado's experiments on brain implants and brain control (Delgado 1969) and Walter Penfield's operations (Penfield 1959, 1975) to remove parts of the brain in epileptic patients.

Though Fuller refers to Milgram with respect of "empathetic cruelty," it is surprising that sociologist Fuller does not discuss the proactive nature of social science experiments such as Zimbardo's "Stanford Prison Experiment" or Milgram's "Obedience to Authority" experiment (Milgram 1974; Zimbardo 2008). A major reaction to these experiments among social psychologists was not to wonder what the experiments revealed, but to provide ethical objections to the experiments. Yet Zimbardo's and Milgram's social psychology experiments became widely popularized and shocked the public about the ease with which normal, average people could become sadists in the right social setting. These experiments explained how Nazi concentration camp guards, committers of war crimes in Vietnam, and torturers at Abu Ghraib in Iraq need not be monsters and could be perfectly normal people before they were set into positions of extreme authority and/or made obedient to extreme authority themselves.

The Nazi hypothermia experiments, despite the unwillingness of many to make use of the results of these murderous experiments, produced genuine data of use to preparing pilots who might be drowned in arctic waters (Caplan 1992; Berger 1990), but many other Nazi "experiments" had no knowledge payoff. Thus, we view that some sort of mechanism for the protection of human subjects should be in place, even if it is a much more liberal one than presently exists. Such a policy, while more attuned to the benefits of research on human subjects, should make sure that subjects willingly participate in projects involving risks and be promised as well as receive full indemnification if negative results occur (Remedios and Dusek 2016).

Expertise

Expertise is highly valued in science, the justice system, in universities, and among professionals in Western-influenced societies. The normative implications are different between object-oriented and agent-oriented social epistemology on expertise. There are strong normative implications with knowledge divided into domains with epistemic agents who rely on experts in each of those domains for object-oriented social epistemology. The objects of knowledge dictate what needs to be known, though epistemic agents rely on experts who analyze those domains and authorizes the goals of knowledge for object-oriented social epistemology. Virtues of knowledge expertise with trust of and deference to authority are emphasized in analytic social epistemology.

As anti-expert, in agent-oriented social epistemology, the epistemic agent seeks to integrate different domains of knowledge into her or his own aims and goals. Fuller recommends agent-oriented social epistemology be critical of purported experts with different normative implications for the agent-oriented social epistemologist, who evaluates knowledge domains based on her or his own aims. Contrasted to this view is object-oriented social epistemology in which the expert is relied on to dictate the ends of knowledge. For instance, Catherine Elgin's epistemic agent uses Putnam's notions of "division of linguistic labor" and obedience to (epistemic) authority, trusting the experts. Just as the utopian socialists and the early Marx criticized the division of labor in economics, so Fuller's notion of the epistemic agent rejects the uncritical acceptance of the division of intellectual labor.

There are many cases where excessive reliance on the division of intellectual labor and trust in experts of object-oriented social epistemology seem appropriate. In recent years both social scientists and lay people have become more aware of the economic and political issues that have affected the deliverances of scientific experts or specialists in various fields. Some of the most egregious cases have been in medical testing, where prestigious medical doctors have accepted pay for signing their names to favorable articles written by employees of the pharmaceutical industry hoping to market the drug. Less egregious, but still misleading, are articles favoring the safety of a drug authored by medical doctors with an interest in marketing that drug (Krimsky 2004). Fudging and massaging of data also occur in military weapons testing (Taubes 2002) and in the process of peer review, grant awards, and other support of publications favoring other politically influential scientists' research programs (Bell 1992). "Experts" with conflicts of interest involving their paymasters or their own research agendas are, increasingly often, not expert sources of knowledge. The subject-oriented epistemological agent therefore needs to participate in the evaluation of the material provided by the alleged expert.

It is not so easy to separate "the scientists'" pronouncements on their specialties from the more or policy-driven, ideological, or anti-religion pronouncements. The object-oriented epistemological agent therefore needs to participate in the evaluation of the material provided by the alleged expert.

Edward Teller was undoubtedly the authority on the physics of hydrogen bombs (after all he invented them), but his claims about the benefits of nuclear testing, for instance, and even the benefits of mutations arising from them, though invested with authority, were not part of the his core expertise. Linus Pauling, similarly, used his expertise in quantum and organic chemistry to deny the safety of nuclear testing, emphasizing its effects on the body (Gorley and Fellet 2015).

A particularly egregious misuse of supposed expertise is when some scientists and philosophers use their prestige as experts on biology to pronounce their religious or non-religious views. The "new atheists" such as Richard Dawkins claim that atheistic conclusions derive from Darwinian biology. Philosopher of biology Michael Ruse outraged Daniel Dennett by saying that in teaching atheism as part of Darwinist science, "they are liable to be faulted for teaching religion (atheism) in public schools." Dennett replied by excommunicating Ruse from the community of respectable philosophers of biology. Dawkins and the historian of population genetics William Provine claimed that their assertion of the truth of atheism is part of their science, while others would strongly disagree and claim the "new atheists" are simply foisting their anti-religious beliefs as science (Brown 2006).

Many scientists have debunked and ridiculed various informal claims of cancer among children, so-called cancer clusters. The experts claimed these clusters were random statistical flukes. The high cancer rate at Love Canal was first brought to general public attention not by scientific experts, but by housewives (Gibbs 2011; Levine 1982).

Another area in which deference to expertise and specialists is undesirable is in hereditary IQ studies. The "experts" in behavioral genetics and IQ studies are almost all strongly committed to hereditarianism (Snyderman and Rothman 1988). The critics of hereditarian theories of IQ are people such as Noam Chomsky, Steve Gould, Richard Lewontin, David Layzer, and Ned Block (Block and Dworkin 1976). This group includes a formal linguist, two evolutionary biologists, a physicist, and a philosopher. These brilliant mostly highly mathematically knowledgeable critics are not members of the professional community of IQ heredity scholars. The one scientist of the above group who was not a mathematical virtuoso, Steve Gould, often dismissed by Fuller as a mere natural historian, made an extensive study of analysis of variation to make his criticisms. The hereditary IQ studies community almost uniformly support hereditarianism, and in a not unsubstantial number of cases, advocated so-called race realism, or what used to be called "scientific racism." The mostly hereditarian "professionals" can dismiss the critics as not being "specialists," credentialed, or experts in the field.

Fuller's social epistemology has a number of advantages over analytic social epistemology. First, the epistemic agent makes knowledge to construct reality instead of the agent justifying knowledge as in analytic social epistemology. Second, Fuller allows for changes in the knower based on converging technosciences rather than assuming a constant nature of the knower. Third, Fuller's account considers corporate and communal knowers in addition to individual knowers. Fourth, Fuller is more critical and reflective about obedience to alleged scientific expertise than its analytic alternative. Finally, Fuller's account encourages a proactive rather than a precautionary approach allowing more daring scientific investigation of human behavior and physiological structures.

Though we support the advantages of Fuller's notion of epistemic agency over analytic social epistemology's notion of agency, we find fault with some aspects of Fuller's notion of agency that he attributes to the scientist. We agree with Downes (2000) that Fuller appears to apply a thin notion of agency to scientists. We note that Fuller's notion of "scientist," which lacks psychological richness and is associated with a thin notion of agency, seems to be incompatible with Tversky's and Kahneman's experimental results (Tversky and Kahneman 1974, 1983) which seem to favor a thick notion of agency. They contend that humans making rapid decisions do not reason according to the rules of probability and decision theory. As a result of their experimental work, Kahneman and Tversky introduce agents who are psychologically rich.

The agents' reasoning can be biased, and, rather than being guided by rational choice theory, the agents' reasoning is guided by heuristics, which are rules of thumb and mental shortcuts for problem solving.

Fuller notes that his thin notion of "scientist" is not a rational utility maximizer but is someone who, in her or his technical role, is neither professionally trained nor personally motivated to reflect on her or his practices as the scientist as epistemic agent is making knowledge through experimentation. Fuller argues that the thick notion of "scientist" excludes the notion of "scientist" as one who is trained to have a "scientific mindset." Fuller, ironically, notes that if the thick notion is applied to the notion of "scientist," then that application eliminates the notion of "scientist" altogether. Fuller's point is that philosophers and psychologists of science have taken "scientist" *qua* "scientist" to mean someone who is rational and lacks psychological richness, and hence, he holds that a thin notion of agency is applicable. Furthermore, Fuller holds that the difference between his notion of a scientist and his notion of a knowledge policy analyst is that the knowledge policy analyst is trained to reflect on scientists' practices by reflexively applying scientific understanding to the scientists' behavior, whereas the scientist is not (Fuller 2000). A scientist, however, can become a knowledge policy analyst with training.

Certainly there are a minority of working scientists who do reflect in a disciplined way concerning scientists' behavior. One can say that they are not philosophers or sociologists of science, but are reflecting in an informal and possibly philosophical manner on their activities. For example, there are scientists such as Paul Gross, Norman Levitt, and Alan Sokal who attacked the reflective discourse of sociologists and anthropologists. Other mathematicians and scientists such as Gabriel Stolzenberg, David Mermin, and Jay Labinger have reflected on the social nature of and aims of science, or, at least, their own science. For example, Jay Labinger wrote a study of the history of the status of inorganic chemistry (Labinger 2013).

Clearly some scientists, whatever one may think of their policy views, have also been knowledge policy analysts. Oppenheimer and Teller as dueling government advisors on nuclear weapons policy, James Watson as head of the Human Genome Project, Bernard Davis as the organizer of opposition to it, Teller and Pauling holding opposite views on the benefits and risks of nuclear testing, Warren Weaver playing a major role in the funding of molecular biology (and opposition to British Marxist and organismic biophysicists) (Abir-Am 1983), and Frederick Lindeman during WWII—these were all advisors on knowledge policy as well as sometime scientists (Remedios and Dusek 2016).

Notes

1. Discussed in Remedios (2003a, b) Chapter 5.
2. Hacking has a notion of maker's knowledge in Representing and Intervening (1983), but Hacking is not advocating for transhumanism.

References

Abir-Am, P. (1983). The Discourse of Physical Power and Biological Knowledge in the 1930s: A Reappraisal of the Rockefeller Foundation's 'Policy' in Molecular Biology. *Social Studies of Science, 12*(3), 241–282.

Bell, R. (1992). *Impure Science: Fraud, Compromise, and Political Influence in Scientific Research*. New York: Wiley.

Berger, R. L. (1990). Nazi Science: The Dachau Hypothermia Experiments. *New England Journal of Medicine, 322*(20), 1435–1440.

Block, N., & Dworkin, G. (1976). *The IQ Controversy*. New York: Pantheon.

Brown, A. (2006, March 6). When Evolutionists Attack. *The Guardian*. http://www.theguardian.com/world/2006/mar/06/religion.uk

Callon, M. (1986). Some Elements of a Sociology of Translation: Domestication of the Scallops and the Fishermen of St Brieuc Bay. In J. Law (Ed.), *Power, Action and Belief: A New Sociology of Knowledge* (pp. 196–233). London: Routledge & Kegan Paul.

Caplan, A. (Ed.). (1992). *When Medicine Went Wrong: Bioethics and the Holocaust*. Totowa: Humana Press.

Delgado, J. M. (1969). *Physical Control of the Mind: Toward a Psychocivilized Society*. New York: Harper.

Downes, S. (2000). Review of *Science* by Steve Fuller. *Philosophy of the Social Sciences, 30*(1), 140–145.

Elgin, C. (2013). Epistemic Agency. *Theory and Research in Education, 11*(2), 135–152.

Fuller, S. (2000, February 22). E-mail to Remedios.

Fuller, S. (2013). The Dawn of Critical Neuroscience. *History of the Hunab Sciences, 26*(3), 107–115.

Fuller, S. (2014). *Knowledge: The Philosophical Quest in History*. New York: Routledge.

Fuller, S. (2016a). *The Academic Caesar: University Leadership Is Hard*. London: Sage Publications.

Fuller, S. (2016b). A Sense of Epistemic Agency Fit for Social Epistemology. In P. Reider (Ed.), *Social Epistemology and Epistemic Agency* (pp. 21–42). London: Rowman International.

Gibbs, M. L. (2011). *Love Canal: The Story Continues*. Washington: Island Press.

Gorley, M., & Fellet, M. F. (2015). The Pauling-Teller Debate: A Tangle of Expertise and Values. *Issues in Science and Technology, 31*(4, Summer), 78–82.

Hacking, I. (1983). *Representing and Intervening*. Cambridge: Cambridge University Press.

Krimsky, S. (2004). *Science in the Private Interest: Has Corrupted Biomedical Research?* Lanham, MD: Rowan and Littlefield.

Labinger, J. (2013). *Up from Generality: How Inorganic Chemistry Finally Became a Respectable Field*. New York: Springer Verlag.

Latour, B. (1987). *Science in Action, How to Follow Scientists and Engineers Through Society*. Cambridge, MA: Harvard University Press.

Levine, A. G. (1982). *Love Canal: Science, Politics, People*. Lanham, MD.: Rowan and Littlefield.

Lynch, W. (2001). *Solomon's Child*. Stanford: Stanford University Press.

Lynch, W. (2016). Social Epistemology Transformed: Steve Fuller's Account of Knowledge as a Divine Spark for Human Domination. *Social Epistemology Review and Reply Collective*.https://social-epistemology.com/2016/05/26/social-epistemology-transformed-steve-fullers-account-of-knowledge-as-a-divine-spark-for-human-domination-william-t-lynch/

Milgram, S. (1974). *Obedience to Authority: An Experimental View*. New York: Harper & Row.

Peirce, C. S. (1958). Note on the Theory of the Economy of Research. In A. W. Burks (Ed.), *Collected Papers* (Vol. 7, pp. 76–83). Cambridge, MA: Harvard University Press.

Penfield, W. (1959). *Speech and Brain Mechanisms*. Princeton: Princeton University Press.

Penfield, W. (1975). *The Mystery of the Mind*. Princeton: Princeton University Press.

Pérez-Ramos, A. (1988). *Francis Bacon's Idea of Science and the Maker's Knowledge Tradition*. Oxford: Oxford University Press.

Remedios, F. (2003a). *Legitimizing Scientific Knowledge*. Lanham: Lexington Books.

Remedios, F. (2003b). Fuller and Rouse on the Legitimation of Scientific Knowledge. *Philosophy of the Social Sciences, 33*(4), 444–463.

Remedios, F., & Dusek, V. (2016). Fuller's Social Epistemology and Epistemic Agency. In P. J. Reider (Ed.), *Social Epistemology and Epistemic Agency* (pp. 61–74). London: Rowman and Littlefield International.

Rescher, N. (1989). *Cognitive Economy: Economic Perspectives in the Theory of Knowledge*. Pittsburgh: University Pittsburgh Press.

Roco, M. C., & Bainbridge, W. S. (2003). *Converging Technologies for Improving Human Performance*. Dordrecht: Kluwer.

Rouse, J. (1996). *Engaging Science: How to Understand Its Practices Philosophically*. Ithaca: Cornell University Press.

Rouse, J. (2002). *How Scientific Practices Matter: Reclaiming Philosophical Naturalism*. Chicago: University of Chicago Press.

Rouse, J. (2003). Remedios and Fuller on Normativity and Science. *Philosophy of the Social Sciences, 33*, 464–471.

Schmidt, J. C. (2011). Toward an Epistemology of Nano-technosciences: Probing Technoscience from a Historical Perspective: On Today's Surprising Prevalence and Relevance of Francis Bacon. *Poiesis & Praxis, 8*, 103–124. https://doi.org/10.1007/s10202-011-0104-z.

Snyderman, M., & Rothman, S. (1988). *The IQ Controversy: The Media and Public Policy*. New Brunswick: Transaction Publishers.

Taubes, G. (2002, April). Postol Versus the Pentagon. *Technology Review*. http://www.technologyreview.com/featuredstory/401412/postol-vs-the-pentagon/

Tversky, A., & Kahneman, D. (1974). Judgment Under Uncertainty: Heuristics and Biases. *Science, 185*, 1124–1131.

Tversky, A., & Kahneman, D. (1983). Extensional Versus Intuitive Reasoning: The Conjunction Fallacy in Probabilistic Reasoning. *Psychological Review, 90*, 293–415.

Zimbardo, P. (2008). *The Lucifer Effect: How Good People Turn Evil*. New York: Random House.

CHAPTER 4

The University and Interdisciplinarity

Abstract This chapter is on Fuller's view that the university is the premier site of knowledge production for the public good. Fuller defends the university against the impact of neoliberalism in which clients influence how academic knowledge is produced. In this context "interdisciplinarity" becomes a battleground. Fuller prefers a version of interdisciplinarity that is regenerative of the university in which academics reach beyond their own fields to a neoliberal version of interdisciplinarity in which academics work in teams for clients on projects. With agent-oriented social epistemology, Fuller's view of interdisciplinarity is that it is internal to agent to organize the disciplines versus object-oriented social epistemology in which disciplines are organized externally by experts. Fuller's view of interdisciplinarity is contrasted to Frodeman's view of transdisciplinarity.

Keywords University • Interdisciplinarity • Transdisciplinarity • Frodeman

Fuller views the university, which is a corporate agent, as the premier site of knowledge production through research for the public good. Fuller defends the university against the impact of neoliberalism in which clients influence how academic knowledge is produced and the state decreases funding. The university is being delegitimized because of neoliberalism and commercialization in which teaching is less valued. Fuller recommends

F.X. Remedios, V. Dusek, *Knowing Humanity in the Social World*,
https://doi.org/10.1057/978-1-137-37490-5_4

the Academic Caesar as the university president to integrate research and teaching. Fuller prefers a version of interdisciplinarity that is regenerative of the university in which academics reach beyond their own fields instead of a neoliberal version of interdisciplinarity in which academics work in teams for clients on projects. This chapter explores the links among the university, interdisciplinarity, and transhumanism. There is a comparison of Fuller on interdisciplinarity and Frodeman on transdisciplinarity on sustainability of knowledge production at the university.

The University and Knowledge Production

For Fuller, the university is premier site of knowledge production and distribution and a solution to the problem of knowledge, which for Fuller is not the epistemological problems of the external world or perception but how knowledge should be organized normatively with a knowledge policy. As a solution to the problem of knowledge, Fuller argues that the outside-in strategy, which is to get from our common reality to the mind-sets of separate people, is the correct strategy instead of the inside-out strategy, which is to get out of our individual heads to a common reality. The outside-in strategy captures the sociology of knowledge, and the inside-out strategy captures the tradition that unites Rene Descartes and traditional naturalistic epistemologists. Fuller argues that most contemporary epistemologies, which can be either internalist or externalist, operate within the inside-out strategy: knowledge is a problem for each individual to solve on his or her terms by a standard to which the cognitive agent may or may not have conscious access. With the inside-out strategy, Fuller avers that there is no sense that epistemic access is a scarce good in which one agent's access to knowledge is impeding, competing with, or making demands on the access of some other agent or agents (Fuller 1996, p. 37). Defenders of the inside-out strategy may argue that Fuller misrepresents their position that epistemic access is a scarce good since they hold that the problem of epistemic access is an epistemic issue. We argue that Fuller's point is that the problem of epistemic access is an economic problem of scarcity, rather than an epistemic problem, for Fuller holds that knowledge is embodied. Fuller holds that epistemic access considered as a scarce good would be in accord with the outside-in strategy, which is typically found in problems posed by philosophers of science; and the portrayal of the cognitive agent is that she chooses with limited resources among research trajectories (Fuller 1996, p. 37; Remedios 2003a, b).

Knowledge, for Fuller is a commodity, which is descriptively a positional good and normatively a public good. The descriptive notion that knowledge is a positional good has value only in terms of its scarcity or restriction on its access, either through academic credentials or intellectual property rights, Fuller holds (Fuller 2002, pp. xviii–xix; Remedios 2003a, b). Though knowledge produced from universities has been espoused to be a public good, which is one that costs less to provide for free than to restrict access, Fuller's view is that because access to academic knowledge is restricted to those who can read published academic journals, academic knowledge is a club good as defined by economists. A club good decreases in value if too many people have access to it. Though few academics may recognize that they are producing knowledge as club goods, Fuller's view is that this is their default position (Fuller 2017).

For Fuller, though the university is the premier site for knowledge production, it is also the institution as the creator and destroyer of social capital since it creates social capital of knowledge through research and destroys social capital of knowledge through teaching. Fuller writes that Wilhelm von Humboldt's reinvention of the university in early nineteenth-century Germany provides creative destruction of social capital as its clearest philosophical justification. The ideal of universal knowledge was reformulated to serve the welfare state's dual economic function of subsidizing capitalist production (research) and redistributing its surplus (teaching) over the last century. During this period, universities increased in size and significance, the welfare state's recent devolution has thrown universities into financial and wider institutional uncertainty with decreased funding and greater scrutiny. Fuller views that the recent drive to have universities to be similar to business firms as generators of intellectual property amounts to no less than a campaign of institutional dismemberment, in which the university's research function is severed from the teaching function (Fuller 2009, pp. 4–5).

With neoliberalism (discussed in Commercializtion section), university research is more client oriented, and research and teaching are being pulled apart. Fuller recommends that universities return to the Humboldtian view, which integrates research and teaching functions. To achieve the combination of research and teaching at the university, Fuller (2016, 2017) recommends the Academic Caesar, as the university president, to take the lead (Fuller 2016). Also Humboldtian is the view that the state should not intervene in the university's production of knowledge (von Humboldt 1970, p. 244). The university is where knowledge is produced for its own sake through research and researchers should be autonomous to pursue research, which will in the future benefit the state (von Humboldt 1970, p. 246). Criticisms of pursuing knowledge for its own sake so that it will eventually benefit society are that society demands practical results in the pursuit of knowledge especially for the publicly funded universities and theoretically to justify the pursuit of knowledge for its own sake on future benefits to society is a category mistake in that the intrinsic value of the pursuit of knowledge is justified on utilitarian grounds (Holbrook 2013). Fuller's response is that though epistemic ends are in the long run ethical ends for utilitarianism and deontology are different in terms of temporal horizons, but they are not different normative stances. Fuller's view is that it is possible to pursue knowledge for its own sake and to justify this policy based on instrumental goals, which is to benefit society, in that those people who attend university, whether they are 10% or 20% of the population, pursue knowledge for its own sake, have a multiplier effect on society when they act of the knowledge they have learned to benefit society (Fuller 2014a). Those people who become doctors or medical researchers can discover cures for diseases. Those people who become civil engineers can design safe roads and dams.

University 2.0

Fuller notes that for Humboldt, who followed Kant, philosophy at the university is the discipline where students resolve which of different disciplines seeking to impose authority for themselves. At the Humboldtian university, the student with an independent and critical intellect is supposed to reach his or her full potential. With humanity 2.0 and transhumanism, where personhood boundaries are disappearing, what is the role of university 2.0? Kurzweil's Singularity University has provided a model, but it is one dimensional in terms of increasing computational performance which is tendentiously linked to rationality and consciousness. But there is a lack of integration of the capacities to the self. Fuller's view is the university 2.0 should follow Humboldt's model to promote *bildung*, which is education where the individual's mind and heart matures in a unification of selfhood and identity in broader society.

Fuller has made the distinction between the proactionary and precautionary principles in which the proactionary principle promotes risk-taking and the precautionary principle promotes risk aversion. On university 2.0, Fuller views that the Academic Caesar can bring proaction to the university by refusing to accept discipline boundaries as rigid and by promoting interdisciplinarity (see Interdiscplinarity section). This is opposed to the neoliberal administrator, who promotes precaution by keeping boundaries rigid and proprietary, a form of epistemic rent-seeking, where rentiers are obliged to site within the rentier's field (Fuller 2016).

The Commercialization of Scientific Knowledge

As knowledge is a commodity, Mirowski and Sent (2007) have highlighted the problem of the commercialization and privatization of scientific knowledge as a positional good. Mirowski argues that the heart of the problem of the commercialization of scientific knowledge is that it is being transformed from a public good to a positional good. Furthermore, Mirowski avers that the commercialization of scientific knowledge is the apotheosis of a neoliberal program to promote and to construct free markets as the central condition of success of the neoliberal agenda (Mirowski 2009, pp. 1–40). A further argument from Mirowski is that science has been harmed by its commercialization (Mirowski 2009, pp. 259–314).

For Fuller, neoliberalism has impacted the university, which is premier site of knowledge production. Clients influence how academic knowledge is produced. Guston has raised the point that during the Cold War, the US government funded 70% of scientific research and private industry funded 30% at universities. Today, the funding has been reversed so the private industry funds 70% and the US government funds 30% (Guston 2002, p. 199).

Fuller provides recommendations for a redesigned institutional structure for the governance of science. The heart of Fuller's proposal is the secularization of science, in which basic research is institutionally separated from the state, and scientific knowledge does not have privileged authority within politics and the public sphere. This leads to a two-stage model for the reorganization of science that parallels Fuller's reinvention of the context of discovery and the context of justification (Fuller 2000). We suggest this is where Fuller addresses the neoliberal notion of science as the marketplace of ideas. The first stage of the model, which parallels the context of discovery, is that the state should stay out of basic research and discovery should be privately funded. This stage leads to a market system of corporations, interest groups, and so on that work to discover knowledge. This system would inevitably produce less new knowledge and would produce fewer unread publications. Hence, science is commercialized and globalized. In the second stage of the model, which parallels the context of justification, state-funded universities would test and distribute privately generated knowledge. This is how commercialized and globalized scientific knowledge is justified and its distribution managed. Scientific knowledge is transformed from a positional good, which is research done by private corporations, to a public good, which is distributed by universities (Remedios 2009).

Interdisciplinarity

In addition to Fuller's recommendations on the governance of science to address neoliberalism, these sections are on interdisciplinarity as Fuller's social epistemology is interdisciplinary because the fundamental question of how knowledge should be organized also asks how can disciplines' boundaries be broken down to improve knowledge production. Fuller views boundaries among disciplines as conventional and institutionalized within the university. Unlike the common view of interdisciplinarity as more cooperation among disciplines, Fuller views that disciplines are artificial holding patterns and discipline success is due to institutionalization. Fuller's social epistemology seeks to break boundaries down among disciplines with interpenetration by rhetoric through science and technology studies. Fuller views that interdisciplinarity is the driver of epistemic change. Compared to conventional science policy, which is problem-centered and does not evaluate the discipline-based knowledge needed to solve a problem, social epistemology's knowledge policy examines the maintenance of institutional inertia: why do research priorities do not change more often? Why do problems arise in some contexts more often than others (Fuller 2007, p. 79).

Fuller's view of interdisciplinarity is that it is more than cooperation among disciplines. For Fuller, at the beginning of modern universities in the nineteenth century, knowledge was unified. From that standpoint, knowledge production was presumed to be "always already" interdisciplinary. Disciplines as we currently know them—corresponding to departments, journals, and dedicated graduate degree programs—only gradually emerged as institutionalized settlements between clashing research programs governed by overarching world views. Thus, people who we now so clearly call physicists, chemists, biologists, physicians, and even engineers were quite hard to distinguish for most of the nineteenth century (Fuller 2009, p. 24).

For Fuller, interdisciplinarity is to recover the lost sense of unity of knowledge. Fuller distinguishes between normal and deviant interdisciplinarity:

> In contrast [to normal interdisciplinarity], deviant interdisciplinarity is more concerned with finding more efficient means of reaching comparable or superior epistemic ends that would allow, at the limit, anyone to know everything. The foundational character of philosophy in the modern Humboldtian university derives from this latter image, an image most imaginatively developed by the German idealists following Kant. (Fuller 2010)

Normal interdisciplinarity views that knowledge was not unified before. Fuller views that interdisciplinarity is reflected in debate between Masters and Doctors in medieval universities. Masters were Franciscans, represented by William of Ockham, who viewed knowledge as organized into university disciplines such as rhetoric, logic, and law are integrated in the individual. Fuller considers that masters are deviant interdisciplinarian agents. Doctors were Dominicans represented by Thomas Aquinas who viewed knowledge is organized by mapping of disciplines of knowledge. Fuller considers doctors to be normal interdisciplinarians. Fuller, who follows Humboldt, views that knowledge integration is achieved by the individual at the university so that the individual becomes self-legislating through education. The social epistemologist is the deviant interdisciplinarian agent, who organizes the disciplines within him or herself like a techno-Goethe (Fuller 2014b; Remedios 2015). This opposed to the normal interdisciplinary agent, who organizes disciplines through mapping them as objects of knowledge (See Chap. 3 on agency). Fuller views that the Academic Caesar can break down discipline boundaries and epistemic rent-seeking with interdisciplinarity (Fuller 2016, p. 102).

How does the university's interdisciplinarity linked to transhumanism? Kurzweil claims that human mind and capacities can be uploaded into computers with increase in computing power. The problem is integration of those capacities and personal identity. Kurzweil's Singularity University has not been able to address the problem of integration. Fuller proposes transhumanities promoted by university 2.0 for integration by the transhumanist.

Examples of Interdisciplinary Research

Fuller contrasts client-driven interdisciplinarity with an intellectually creative interdisciplinary produced by free inquiry in the university. Fuller is justifiably concerned that client-driven interdisciplinarity, a product of neoliberal society, and the fiscal crisis of the state are defunding universities as well as welfare, public health, and other programs.

Fuller also advocates a science oriented toward public participation and justification in terms of service to the community. Both the science oriented toward a republican society and the client-centered science driven by corporate, military, and government demands contrast with absolutely "pure knowledge" or "science for science's sake." Both include awareness of possible social benefits, whether to the whole populace in Fuller's ideal science or to various powers that be in client-driven science.

Even within interdisciplinarity unhindered by corporate client demands, there can be conflicts between an interdisciplinary science oriented toward fundamental theoretical understanding and one more concerned with social usefulness rather than with finding ultimate laws or rock bottom mechanisms. Fuller is sympathetic to the Newtonian ideal of deductive and predictive explanation and contrasts this favorably with the family-of-models approach more recently introduced in philosophy of science in part to account for evolutionary explanation. Nevertheless, Fuller supports a unity of science ideal, not based on reduction, but rather on a unified language and, if possible, a phenomenological, perceptible, approach of the sort favored, though presented in very different ways, by Ernst Mach and Johann von Goethe. This approach would make science more accessible to the public, avoiding theoretical entities and advanced mathematics. This would seem to be in part behind Fuller and Collier's claims that the difficulties of interdisciplinary collaboration are more a matter of rhetorical communication than conceptual (Fuller and Collier 2004). This earlier work on interdisciplinarity can be linked with Fuller's conception of unified science (see Chap. 2), involving not primarily reduction to unobservable micro-enities, but a unified language of science that Mach and the early logical positivists advocated. To achieve such a unified language in discussions between scientists from different specialties with their own jargon rhetorical means may be applied to achieve cooperation and a common language.

Having not only a common language but a language comprehensible to non-specialists and the general public fits with Fuller's goal of the democratization of science. (One problem with this is the role of advanced mathematics in various sciences. Some of Fuller's remarks suggest he would like to eliminate this mathematical technicality.) An advantage of presenting science in a generally comprehensible common language is the conversion of research into teaching, destroying mystification by means of specialized jargon. This is the "creative destruction" of intellectual capital.

Fuller and Collier use as examples of the great innovations of Newton, Darwin, and Freud. These are hugely comprehensive theories unifying various fields, the unification by Newton of physics of earth and the heavens; by Darwin of distribution of organisms, embryology, comparative anatomy, and fossils; and by Freud of psychology, psychotherapy, anthropology, and most of the social sciences (Fuller and Collier 2004, p. 37). These are hardly typical interdisciplinary activities. They do more than link previously existing disciplines. They disrupt and create disciplines.

There are possible conflicts between Fuller's various desiderata for interdisciplinary research and theories. Warnings about the dangers of research driven by social goals are often issued by defenders of "pure" research, often augmented by references to Soviet Stalinist of German Nazi science. This often involves "pure" scientists' professional apologetics and obliviousness to the social and cultural background directing their research choices. Nevertheless, there is some truth in these warnings, whether client driven or dedicated to the well-being of society, the demands on the science can misdirect or distort it. Also, there is a tradeoff between the short-run benefits of phenomenological and utilitarian research and the long-term benefits of research into fundamental mechanisms.

Fuller advocates an interdisciplinarity that does not simply eclectically combine methods and theories of disciplines but that disrupts the original disciplines. However, clearly not all new interdisciplinary interventions of this sort are successful, just as most proposed scientific theories are not successful.

The sort of deviant interdisciplinarity that Fuller advocates disrupts the disciplines rather than simply combining different already existing disciplines and people from the traditional, established disciplines. Molecular biology is a paradigmatic example of this. A striking example of resistance from a traditional discipline is Erwin Chargaff's remark after meeting Watson and Crick, who seemed to him like clowns, ignorant of the component bases that Chargaff had spent his life studying, but beat everyone to the punch on DNA structure. Chargaff said that molecular biology is "practicing biochemistry without a license" and "Pop biochemistry" (Chargaff 1980). Steven Rose writes that this antagonism between biochemistry and molecular biology lasted well beyond Chargaff's encounter with Watson and Crick. Ernst Chain refused to appoint molecular biologists to the biochemistry department of Imperial College. Later Watson had a negative attitude toward the biochemical approach of Craig Venter when both were leaders in the race to sequence the human genome (Rose 2011).

Nancy Maull and Lindley Darden have characterized "interfield theories" as theories that cross and relate fields rather than other theories. These do not relate theories by vertical, hierarchical reduction, but by horizontal unification of language. They call this "unifying science without reduction." The relation is "interfield" rather than interlevel, as it is in reduction. Interfield theories often operate on several levels simultaneously. The alternative to derivational reduction begins by drawing attention to the way a vocabulary can be "shared" by different areas of research. Such "shared" vocabulary, it turns out, can be used to identify a very special kind of problem, a problem that, although it arises from one branch of inquiry, can only be solved with the aid of another science (Darden and Maull 1977, Darden 2005). However, they reject the relevance of sociological analysis to their account. "We are not interested in conceptual, not sociological or institutional change" (ibid., p. 44). They do, however, include historical change of problems and fields. A true account of the nature of interdisiplinarity should include the social organization of the relevant personnel as well as the broader social and historical context.

One advantage of Darden and Maull's account is in dealing with the so-called rules of correspondence where they function to connect the terminology of one theory with another. The links are called "bridge rules," "dictionary," and other things, the multiplicity of terms for this relation showing the term's lack of use by scientists themselves. Darden (2005) treats what reductionists call the rules of correspondence as emerging out of the process of discovery and the research strategies of the relating and integrating different fields of science.

This model of interdisciplinarity fits with Fuller's model for unified science rather than a deduction from a higher-level theory. On the other hand, it may seem too close to the interdisciplinarity involving the cooperation of established disciplines that Fuller opposes. Actually, Fuller's main early example of creative and disruptive interdisciplinarity is Newton's mechanics. Ironically, this unifies fields such as astronomy and mechanics, earth and heavens by means of reductionist deduction. William Whewell's notion of consilience on some interpretations would fit with this. Those who do treat Darwin's theory (earlier used but now rejected by Fuller) as a higher-level theory, natural selection, which explains distribution, embryology, comparative anatomy, and so on. They were not simply a common language.

Cell biology, modern genetics, and molecular biology, the main examples in Darden and Maull's work, are fields praised by Fuller and contrasted by him with the supposed "natural history approach" of evolutionary biology. They certainly draw on previously established disciplines such as physics, genetics, and biochemistry, even if practicing them "without a license." The same goes for neuropsychology and cognitive science.

Some of the novel theories have been founded by creative individuals rather than teams of researchers from other disciplines and are not produced by cooperation of individuals from different specialties. These include the paradigmatic examples of Newton's, Darwin's, and Freud's theories. Others are Vernadsky's biogeochemistry, Richard Levins' "Evolution in Changing Environments" (1968), combining the dynamics of both evolution and ecology, and Watson and Crick, if we count the dynamic duo of DNA fame as a "Janus headed creature," as they have been called, rather than separate people from different fields.

One of the ways that interdisciplinary research can unify the research and teaching unity of Humboldt's and Fuller's ideal is doing research done with lower tech and less expensive apparatus that is accessible to the students. To generalize this approach would involve downgrading giant particle accelerators or multinational astronomical measuring devices to the use of more student-accessible equipment. This has occurred in recent decades by biology replacing physics as the focus of novel and applicable research. The first gene spicing by Hamilton Smith in the 1970s was done in sophisticated professional labs but now can be done in high school or by using a kitchen sink. Philosopher of science Paul Feyerabend was inspired to an interest in public, participatory science when as a teenager he gave astronomy lectures for the mass gatherings of star gazers in Vienna convened by Paul Thomas (Feyerabend, p. 28). In England in the 1930s and 1940s, the Mass-Observation movement enlisted groups of 500 citizens to record opinions and observations. Despite criticisms by quantitative social scientists, the results were used in government policies as well as by the economist Lord John Maynard Keynes (Hinton 2013).

An example of interdisciplinary teaching of a new discipline that is directly tied to teaching is Big History. This integrates cosmology, geology, evolutionary biology, archaeology, and traditional history, the later including natural resources (Speier 2015). Bill Gates has supported this movement, supplying high-quality websites and teaching materials at no cost, spreading the movement into secondary schools as well as colleges. Another kind of interdisciplinary teaching can link science with the arts and humanistic studies.

There are numerous courses linking environmental science with literature about the land and nature writing. There are even more integrative and interdisciplinary courses of science and the visual arts, which have been taught at the National University of Singapore, Dartmouth College, Smith College, University of New Hampshire, and elsewhere. These integrate notions of symmetry in physics, chemistry, and mineralogy with symmetry in the arts and crafts. This can include Renaissance art and twentieth-century abstract art. One topic is Renaissance perspective painting and its historical influence on mathematics, science, and philosophy. These courses can draw on works by mathematicians (Grünbaum, Schattschneider, Senechal), art historians (Edgerton, Ivins), and philosopher (Harries) to link science and art.

Courses on video art can also tie in science. Other courses integrate color science, both physical and psychological, with literature and art (Gage, Gass, Riley) in the manner of Goethe's theory of colors (Goethe). Less frequently taught courses integrate music with science, linking classical music with physics (Levenson, James) or electronic music (Pierce). These courses can "humanize the scientist and Simoniz the humanist."

There have been first year cross-department college courses centered on around a single drama that links mathematics, art, economics, and/or religion. One such course used Tom Stoppard's *Arcadia* (Stoppard) tying together chaos theory and computer programming in a fictionalized life of Lord Byron's daughter and, as assistant to Babbage in the early 1800s, the first computer programmer. Another course centered around the movie thriller *Pi*, involving mathematics, the Kabbalah, and the stock market, though the latter might not have been inspiring to all college students, when in the end the exploited hero stops his mathematical obsession by drilling a hole in his brain.

Such interdisciplinary and team-taught courses are threatened by the neoliberal university's emphasis on specialized research and publication, cutbacks on educational resources, and elimination of art and music courses in US secondary education in favor of "readin', writin', and repressin'," testing on language and mathematics leading to the downplaying of courses on art, music, history, social studies, and even science, as they are not included on national tests.

An example of this conflict between contrasting modes of interdisciplinary research is research into biological clocks. Two approaches at the beginning of the discipline were on the one hand the search for mechanisms by Colin Pittendrigh (1960) and Jürgen Aschoff (1960) and on the other an attempt at a general science of chronobiology by Halberg. Pittendrigh and allies saw research into biological clocks as an application of general physiology, cell biology, and biochemistry. Pittendrigh saw no need for a separate discipline of biological cycles. On the other hand, Halberg advocates such a separate discipline studying biological cycles in general. Halberg's approach is oriented toward medical and psychiatric applications. Halberg disparages the fundamental mechanism people as "clock watchers," while Pittendrigh sees no more need for a general science of chronobiology than for a general science of homeostasis, to which Halberg often compares his proposed discipline. While those following the Pittendrigh approach publish in established zoology, botany, and biochemistry journals, Halberg's allies commandeered a previously existing but at the time largely moribund journal and founded a society for chronobiology.

Interestingly, the science studies people who wave written on biological clock research (William Bechtel, Cambrosio, and Keating) have focused their attention and spent almost all their time on the Halberg program. This may be precisely because the Halberg program is both more phenomenological and more applied. It may also because Halberg is still alive and available for lengthy interviews (Halberg, *Journal of Circadian Rythms*, n.d., http://www.jcircadianrhythms.com/articles/10.1186/1740-3391-1-2/), while Pittendrigh, who was a great speaker and raconteur, is deceased. William Bechtel, for instance, though having investigated other more theoretical interdisciplinary programs, such as cell biology (Bechtel 1993), is now focused in the medical and psychiatric applications of biological clock research.

Halberg's program is broader (treating temporal biological phenomena in general) as well as more obviously, directly, tied to applications.

Nevertheless, a general science of cycles risks becoming vacuous. There is a society for cycle research in general, which includes not just biological cycles but economics and other ones. Several societies of this sort lose respectability, not merely because of professional specialization and jealousy, but because the results can become empty and untestable. Deevey, a student of fossil pollen, long ago pointed out that purely phenomenalist application of cycles analyzed in terms of sines and cosines can be used to describe any pattern, and the approach becomes unfalsifiable in a Popperian sense (Deevey 1960). Another pioneer in clock research, Frank A. Brown, claimed to find all sorts of correlations of biological clocks in invertebrates such as horseshoe crabs with all sorts of astronomical and other phenomena, and this part of his labors was soon discredited (Brown 1960; Brown et al. 1970). This is the issue that Paul Feyerabend once pointed out for Ptolemaic astronomy with its epicycles, combining circles rotating about points on circles (Fourier analysis) can account for any curve whatsoever, and hence, the method cannot be refuted.

A recent survey of biological clock research by a major actor in the field has numerous references to Pittendrigh, Menaker, and Aschoff, but though Halberg appears in the bibliography, he does not even appear in the index (Foster and Kretzman 2005).

As it turned out, the apparently less practical and applicable biochemical and cytological approach eventually paid off. Seymour Benzer, who earlier contributed to the development of the transistor and had done an insightful analysis of different meanings of the gene (muton: cistron, and codon: mutation, recombination, and genetic code), found a gene for biological rhythm in the fruit fly (Weiner 1999, pp. 105–111; Foster and Kretzman 2005, p. 98, 102). Michael Menaker located brain mechanisms responsible for bird and mammal circadian rhythms (Menaker 1968) and collaborated on the identification of a clock gene (Lowrey et al. 2000). Another, even more theoretical, and initially much more speculative, theory of biological rhythms was developed by Brian Goodwin (a student of Waddington, who later advocated of a non-selectionist, structuralist account of evolution). This theory involved modeling temporal rhythms of cells on the abstract cells of statistical mechanics, using analogues to thermal physics, such as "temperature," and using non-linear mathematics to model biological clock cycles. This model remained alien to most physiologists until decades later, when it was combined with new anatomical and biochemical knowledge, because it is now applicable to real biological clocks (Edmunds 2012).

This field exemplifies what appear to be conflicts among Fuller's criteria for good interdisciplinarity. The phenomenalistic approach may be more easily teachable, more comprehensible to the public, and more immediately, practically applicable, but the "Newtonian" approach that Fuller elsewhere advocates can conflict with the phenomenological one, especially as applied to atomic explanations. The molecular biology and computer approach to life, which Fuller favorably contrasts with the "natural history" approach, associated by Fuller with evolutionary theory, is itself a non-phenomenological approach.

Transdisciplinarity

In this section, we contrast Fuller's view of interdisciplinarity to Frodeman's view of transdisciplinarity, which is about sustainability of knowledge production at universities and the co-production of knowledge from inside and outside the university. Fuller's view is that knowledge is limitless and Frodeman's view is that knowledge is limited. Frodeman (2013) frames his argument in environmental terms for sustainability of knowledge production. For Frodeman, too much knowledge is not balanced with other qualities such as "courtesy, solidarity, and quietude" (Frodeman 2013, p. 5). Frodeman's view is that knowledge production should be slowed down so that there is progressive change because the public feels alienated from technological change. According to Frodeman, there is too much knowledge produced and the cost of producing so much knowledge is unsustainable. Frodeman diagnoses the problem to be disciplinary knowledge production, which is to seek unlimited knowledge within the discipline's domain. Frodeman's solution is interdisciplinary knowledge production, which is limited to solving a specific problem that individual disciplines cannot solve. Frodeman argues that with today's shrinking budgets, growth of massive online courses (MOOCS), growth of knowledge through Internet, disciplines are losing their authority, universities are losing their role as the premier site of knowledge production and as knowledge producers are becoming unsustainable. Frodeman uses Fuller (2011) as an example of someone who is promoting creation of too much knowledge by linking technology and humanity 2.0's proactionary principle.

Frodeman uses the concept of sustainability with knowledge production at universities. Frodeman narrows his argument that knowledge production at universities is becoming unsustainable. Though universities are facing challenges such as reduced funding, growth of online courses, growth of knowledge on Internet, there is no conclusive evidence that universities are losing their role as the premier site of knowledge production and are becoming unsustainable. There is a problem of conversion of university research to commercialized products but that is not the same problem as universities losing their role as the premier site of knowledge production. Frodeman's argument that universities through disciplinarity are producing too much knowledge, which is unsustainable, is like the argument of the Slow Food movement, which promotes local cuisines and growing methods against too much fast food and globalization of companies like McDonalds. For Frodeman, interdisciplinary research is to solve specific problems, which will make universities sustainable. This notion of sustainability is economic. How will universities survive even though modern universities are founded on the Humboldtian model in which research is unlimited?

For Frodeman, transdisciplinarity is co-production of knowledge from inside and outside the university. Will it be university researchers or clients who set the agenda on which interdisciplinary problems to solve? Modern universities are based on Humboldt's model of research and teaching. Research is the endless search for knowledge. By suggesting limits to research and to knowledge, Frodeman is in effect arguing that Humboldt's model is no longer viable. What will be the distinguishing features of future universities if they do not follow Humboldt's model? Frodeman's response is that his view does not imply the death of the Humboldtian model of the university. Frodeman views that there should be a plurality of models of universities (Frodeman 2014).

Frodeman's argument that Fuller's view to increase knowledge production through linking technology and humanity 2.0's proactionary principle is unsustainable misses Fuller's transhumanism, which is Fuller's basis for the proactionary principle. With humanity 2.0, sustainability is not a problem because silicon-based systems such as androids can have personhood. The difficulty is the transition from humanity 1.0 to 2.0, which has not arrived and the responsibility of humanity 2.0 to 1.0.

Our assessment is Fuller and Frodeman are talking about two different types of interdisciplinarity. Fuller's version of interdisciplinarity is deviant interdisciplinarity where the site of interdisciplinarity is the epistemic agent, who can internally organize the disciplines and unify knowledge (see Chap. 3 on epistemic agency). Frodeman's version of interdisciplinarity is normal interdisciplinarity where the site of interdisciplinarity is at the university among researchers or a group of researchers where the knowledge from disciplines is traded in trading zones.

Furthermore, Frodeman has not grasped Fuller's argument above concerning the Humboldtian university where knowledge is pursued for its own sake and it is limitless, and the justification for this view is on instrumental grounds of future benefit to society. To recap Fuller's argument, epistemic ends are ethical ends in the long run for the normative stances of deontology and utilitarianism are different in degree. Those people who attend university to pursue knowledge for their own sake will benefit society when they act on the knowledge they have learnt in discovery of cures for diseases if they are doctors or medical researchers or build safe roads or dams if they are civil engineers. Fuller's view is that those who graduate from university have a multiplier effect on society.

References

Aschoff, J. (1960). Exogenous and Endogenous Components in Circadian Rhythms. *Cold Spring Harbor Symposium Quantitative Biology, 25*, 11–28.

Bechtel, W. (1993). Integrating Sciences by Creating New Disciplines: The Case of Cell Biology. *Biology and Philosophy, 8*, 277–299.

Brown, F. A., Jr. (1960). Response to Pervasive Geophysical Factors and the Biological Clock Problem. *Cold Spring Harbor Symposia in Quantitative Biology, 25*, 57–71.

Brown, F. A., Woodland Hastings, J., & Palmer, J. D. (1970). *The Biological Clock: Two Views*. New York: Academic Press.

Chargaff, E. (1980). "Gullible's Travels," and "Matches for Hersostratus". In *Heraclitean Fire* (pp. 100–103). New York: Warner Books.

Darden, L. (2005). Relations Among Fields: Mendelian, Cytological and Molecular Mechanisms. *Studies in History and Philosophy of Science Part C: Studies in History and Philosophy of Biological and Biomedical Sciences, 36*(2), 349–371.

Darden, L., & Maull, N. (1977). Interfield Theories. *Philosophy of Science, 44*, 43–64.

Deevey, E. (1960). The Hare and the Haruspex: A Cautionary Tale. *American Scientist, 48*(3), 415–430.

Edmunds, L. N., Jr. (2012). *Cellular and Molecular Bases of Biological Clocks: Models and Mechanisms for Circadian Timekeeping*. New York: Springer.

Foster, R. G., & Kretzman, L. (2005). *Rhythms of Life: The Biological Clocks That Control the Daily Lives of Everything*. New Haven: Yale University Press.

Frodeman, R. (2013). *Sustainable Knowledge: A Theory of Interdisciplinarity*. London: Palgrave.

Frodeman, R. (2014). Sustainable Knowledge: An Exchange, Davis, Dieleman, Frodeman, Remedios, Riggio, Simbürger, Suomela. https://social-epistemology.com/2014/03/15/sustainable-knowledge-an-exchange-davis-dieleman-remedios-riggio-simburger-suomela/#more-5189

Fuller, S. (1996). Social Epistemology and the Recovery of the Normative in the Postepistemic Era. *Journal of Mind and Behavior, 17*(2), 83–97.

Fuller, S. (2000, February 22). E-mail to Remedios.

Fuller, S. (2002). *Social Epistemology* (2nd ed.). Bloomington: Indiana University Press.

Fuller, S. (2007). *New Frontiers in Science and Technology Studies.* New York: Polity Press.

Fuller, S. (2009). *The Sociology of Intellectual Life.* London: Sage Publications.

Fuller, S. (2010). *Science.* Durham: Acumen Publishing Limited.

Fuller, S. (2011). *Humanity 2.0: What It Means to Be Human Past, Present and Future.* London: Palgrave Macmillan.

Fuller, S. (2014a). Comments to J. Britt Holbrook 'Fuller's Categorical Imperative: The Will to Proaction'. *Social Epistemology Review and Reply Collective, 2*(11), 20–26. http://wp.me/p1Bfg0-13K

Fuller, S. (2014b). Social Epistemology: The Future of an Unfulfilled Promise. *Social Epistemology Review and Reply Collective, 3*(7), 29–37. http://wp.me/p1Bfg0-1wG

Fuller, S. (2016). *The Academic Caesar: University Leadership Is Hard.* London: Sage Publications.

Fuller, S (2017). The Case for Academic Caesars. *Times Higher Education Magazine*.https://www.timeshighereducation.com/features/case-academic-caesars

Fuller, S., & Collier, J. (2004). *Philosophy, Rhetoric, and the End of Knowledge: A New Beginning for Science and Technology Studies* (2nd ed.). Mahwah: Lawrence Erlbaum Associates, Publisher.

Guston, D. H. (2002). Secularising Science? *Futures, 34*, 197–199.

Hinton, P. (2013). *The Mass Observers: A History, 1937–1949.* Oxford: Oxford University Press.

Holbrook, J. B. (2013). Fuller's Categorical Imperative: The Will to Proaction. *Social Epistemology Review and Reply Collective, 2*(11), 20–26. http://wp.me/p1Bfg0-13K

Levins, R. (1968). *Evolution in Changing Environments.* Princeton: Princeton University Press.

Lowrey, P. L., Menaker, M., et al. (2000). Positional Synthenic Cloning and Functional Characterization of the Mammalian Circadian Mutation tau. *Science, 288*(5465), 484–491.

Menaker, M. (1968). Extraretinal Light Perception in Sparrow, I. Entrainment of the Biological Clock. *Proceedings of the National Academy of Sciences USA, 59*(2), 414–421.

Mirowski, P. (2009). *Science-Mart*. Harvard: Harvard University Press.
Mirowski, P., & Sent, E.-M. (2007). The Commercialization of Science, and the Response of STS. In E. J. Hackett, O. Amsterdamska, M. Lynch, & J. Wajcman (Eds.), *The Handbook of Science & Technology Studies* (3rd ed., pp. 635–690). Cambridge, MA: MIT Press.
Pittendrigh, C. S. (1960). Circadian Rhythms and Circadian Organization of Living Systems. *Cold Spring Harbor Symposia in Quantitative Biology, 24*, 159–184.
Remedios, F. (2003a). *Legitimizing Scientific Knowledge*. Lanham: Lexington Books.
Remedios, F. (2003b). Fuller and Rouse on the Legitimation of Scientific Knowledge. *Philosophy of the Social Sciences, 33*(4), 444–463.
Remedios, F. (2009). Fuller and Mirowski on the Commercialization of Scientific Knowledge. In J. van Bouwel (Ed.), *The Social Sciences and Democracy*. Basingstoke: Palgrave Macmillan.
Remedios, F. (2015). Knowing Humanity in the Social World: A Social Epistemology Collective Vision? In J. Collier (Ed.), *The Future of Social Epistemology: A Collective Vision* (pp. 21–28). London: Rowman and Littlefield International.
Rose, S. (2011). Practicing Biochemistry Without a License. *EMBO Reports, 12*(5), 381.
Speier, F. (2015). *Big History and the Future of Humanity*. Chichester: Wiley.
von Humboldt, W. (1970). On the Spirit and the Organizational Framework of Intellectual Institutions in Berlin. *Minerva, 8*, 242–250.
Weiner, J. (1999). *Time, Love, and Memory: A Great Biologist and His Quest for the Origins of Behavior*. New York: Alfred A. Knopf.

CHAPTER 5

Fuller's Intelligent Design

Abstract This chapter is Fuller's version of Intelligent Design (ID) in opposition to evolutionary theorists, and the controversy concerning it is discussed. This chapter discusses why ID is important to Fuller in terms of defending the spiritual distinctiveness of humanity. He claims doing science is to participate in the mind of God, since humanity is created in the image and likeness of God. Fuller contrasts what he claims is the natural history approach of Darwin and evolutionists with molecular biology, claiming ID is linked to the latter. Fuller's claims concerning the scientific status of evolutionary theory are criticized.

Keywords Intelligent Design • Darwin

Intelligent Design (ID) is a recent and controversial theory and movement that rejects evolution. ID differs from earlier creationism in that ID (1) denies the old earth theory based on Biblical literalism and extrapolations from the generations in the Book of Genesis to claim the earth is only thousands of years old and (2) speaks on of God, but of an "Intelligent Designer," vaguely characterized. Mainstream ID refuses to do this, partly because of legal reasons: the banning in the USA of teaching religion in the public (state supported) classroom.

F.X. Remedios, V. Dusek, *Knowing Humanity in the Social World*,
https://doi.org/10.1057/978-1-137-37490-5_5

Fuller's ID is different from that of other deniers of evolution in that he sets his claim amidst general account of the history of science accounts as well as defense of human agency and the capacity of humans to transform into something other than their present biological form (transhumanism). This chapter is on Fuller's version of Intelligent Design (ID) debate with evolutionary theorists and the controversy concerning it. This chapter discusses why ID is important to Fuller in terms of defending the spiritual distinctiveness of humanity, which has epistemic agency and to do science is to participate in the mind of God since humanity is created in the image and likeness of God.

Although Fuller generally seems to accept the old earth of standard science and ID, he at times criticizes radioactive dating of rocks and claims that the reason for mainstream science to hold to an earth billions of years old is in order to give time for Darwin's gradual evolution to work (Fuller 2010, p. 89). In fact, a century before Darwin, non-evolutionist and anti-evolutionary early geologists concluded that the earth is vastly old, hundreds of millions of years, or even infinitely so, based not on evolution, but on examining layers of rock, sedimentation, and other geological phenomena. For instance, the anti-evolutionist Cuvier claimed the earth was thousands of ages old, in which an "age" is some 5000 or 10,000 years, making the earth at least many millions of years old. Adam Sedgwick, who excoriated early writings on evolution and opposed Darwin's views lifelong, introduced the Cambrian epoch on purely geological grounds and believed the earth was extremely old (Albritton, Gillispie, Greene). Even without radiometric dating, rates of deposition of sand earlier gave an age of three million years and in later more recent research an age of one and one half billion years.

Fuller's "Intelligent Design" differs in many respects, both in content and in methods of defense from standard ID. Fuller's rhetoric makes an ingenious contrast between Darwinian evolution, which he associates with natural history, and molecular biology, which he associates with his transhumanism. Later we shall investigate how well this dichotomy holds up. Fuller also is willing in contrast to mainstream ID, in claiming his version of ID should identify God as the Intelligent Designer (Fuller 2009, p. 254).

On his own account, Fuller developed his new direction embracing Intelligent Design, what he calls "Left Creationism" (Fuller 2014c, p. 16) along with transhumanism out of his normative social epistemology (Fuller 2014b). He did this by filling in more specifically the norms which had emphasized as essential to a proper social epistemology emphasized the aspirational nature of humanity with respect not only to social progress and in knowledge of but in the control of nature. Humanity's orientation to knowledge and control of the universe is characterized in terms of the Biblical characterization of humans as image of God (*Imago Dei*). An early portrayal of this quest in intellectual and spiritual terms is Saint Bonaventure's work *The Mind's Road to God* (Bonaventura 1952 [1259]; Fuller 2010, pp. 127–129).

In several articles primarily directed to a Russian audience as well as to Western believers, Fuller has linked the aspirational aspect of humans with respect to the deity to the Eastern Orthodox doctrine of *theosis* of humans, not only striving toward God but also actually participating in God (Fuller 2014c). Fuller's notion of the approach to God seems to go further than even the most extreme of theological treatments of the process. It seems suggested that humans will become God through transhumanist improvements. This promethean goal has led to charges of Gnosticism against Fuller. (See discussion of Gnosticism in Chap. 6.)

In his *The New Sociological Imagination*, Steve Fuller rejects the account of what is essential to human beings in terms of evolutionary biology and Darwinian evolution. Fuller defends the autonomy of "the human" in sociology from biology (Fuller 2006b, pp. 90–107). As with more traditional Biblical literalist creationists, the real concern of his discussion is the status of human beings, their evolution from monkeys, not with the evolution of insects, birds, and other organisms.

In pursuing implications of his normative social epistemology, Fuller turned to the more specific content of the norms and in turn to his treatment of the human beings and the nature of the knowers. Fuller develops the idea that humans as knowers have a special status that is not connected to their biological nature. Fuller's emphasis on the conception of humans as image of God is what leads him to reject Darwin and Darwinism. This leads him to reject biological evolution as such and to support his own version of Intelligent Design (ID). ID claims that living things (and the rest of the universe for that matter) are designed by a Higher Intelligence. Fuller is not coy about the identity of the Intelligence as God as is mainstream ID movement, which, primarily for legal reasons, does not identify the Intelligent Designer as God.

On the relation of humans to other animals and to God, Fuller defends the views of scholastic Duns Scotus rather than those of now official philosopher of Roman Catholicism, St. Thomas Aquinas. Concerning the relation of humans to other animals, Fuller notes that Aquinas closely follows the ancient Greek philosopher Aristotle in designating the human species as a higher form of animal, a "rational animal," but an animal nonetheless. On the relation of humans to God, Fuller notes that Aquinas holds to a qualitative difference between humans and God (Fuller 2011b, p. 80). For Aquinas, there is only an analogical relation between human qualities such as knowledge, goodness, power, and God's knowledge, goodness and power.

In contrast, Duns Scotus, whom Fuller follows, holds that there is only a quantitative difference between humans and God. God is in a sense a vastly, indeed infinitely, boosted human (Fuller 2011b, pp. 78–86). Also because of this godlike mind, humans are qualitatively distinct from other animals.

One aspect of the Franciscan tradition that Fuller does not incorporate is the legends of St. Francis and his regard for animals. St. Francis was named a patron saint of ecology, but Fuller rejects political ecology as involving the precautionary approach toward the environment.

Epicurus and Hume

Epicurus was a philosopher in the second century BCE. He held that the physical universe consisted solely of atoms and empty space. Epicurus emphasized the avoidance of pain, counter to Fuller's proactionary orientation of willing to endanger and sacrifice people for the advance of science.

Epicurus was rediscovered in the Renaissance through the survival of the book by the Roman poet Lucretius' *The Nature of Things* (2008, [94 BCE]). Lucretius presented the views of Epicurus in the form of an epic poem, of which only a few copies survived. Giordano Bruno (who held that there are an infinite universe and world, denying human's special place at the center of the cosmos) was one of the first to use Lucretius. In the seventeenth century, the "century of genius," many scientists and philosophers embraced Epicurean atomism. Of course, they accepted the Christian God, as well as a purely mechanical view of the physical universe. Newton, for instance, accepted atoms, tiny and hard, but had trouble dealing with Epicurus' purely materialistic picture. Newton added "active powers," not mental but not purely material, and human souls. Boyle, Descartes, and many others accepted the atomic view, but in a "theo-mechanical" notion with mechanism, plus minds and God. Fuller links Epicurus' purposeless mechanism and randomness with Darwin.

Fuller rejects Darwin and Darwinism in favor of a very original if idiosyncratic version of Intelligent Design (ID). Besides Darwin Fuller's candidates for *bête noir* are Epicurus and Hume. He calls "Darwin, ever the Epicurean" (2011a, p. 193). Hume was "one of Darwin's youthful inspirations" (Fuller 2008, p. 84).

One can see how the views of Epicurus would be anathema to Fuller in many respects. Epicurus is a total materialist, denying that there is any nonmaterial spirit of humans or of gods. Furthermore, Epicurus denies that there is any higher purpose in the universe. Also, there is no moral interaction of these gods with humans. The one thing that would probably be in his favor for Fuller, though he never says this, is that Epicurus does believe in free will. However, he bases this freedom on randomness in which the collisions and adhesion of atoms account for all the objects and processes in the universe. One peculiarity of Epicurus' views, in contrast to earlier atomists, who held what we would call deterministic views, is that the atoms have "swerve." That is, the atoms randomly depart from their paths. Epicurus uses this to account for free will of humans. There is the existence of (material) gods, but they are totally uninterested in human affairs.

Also, Epicurus, like Darwin, "studiously avoided reaching any firm conclusions concerning the origin, ends, or design of the universe" (Fuller 2010, p. 39). Fuller sees the views of Epicurus as similar to "karmic" Asian thought, linking this with August Weismann's "germ plasm" and genetic determinism (2008, pp. 139–140). Oddly, Fuller generally defends E. O. Wilson and Richard Dawkins, who are closest to genetic determinism, and rejects Stephen Jay Gould and Richard Lewontin, who are opponents of genetic determinism.

Another figure closer to Darwin's day was the eighteenth-century philosopher Hume. Despite Hume's later acceptance by liberal or even social democratic figures such as the logical positivists, Fuller emphasizes how much of a traditional conservative, and indeed racist, were Hume's social views. (Of course, this might seem ad hominem against Hume's views in theory of knowledge and metaphysics, but Hume's appeals to custom as more fundamental than reason do certainly fit with his conservatism.)

Hume claimed that we can know hardly anything on purely rational grounds. There is no good logical reason to support the belief that the future will be like the past (induction). There is furthermore no justification for our belief in an external world and independent physical objects therein. There is not even a reason to believe in a unified self beyond scattered perceptual images and ideas.

On the other hand, outside of rational philosophy, we believe in these things by "custom and habit." Thus, given the weakness of reason, Hume appeals to common practice. Reason is a slave to the passions. This links up with his traditionalist conservatism. If reason is weak, tradition is all that can be trusted.

The logical positivists in the early twentieth century took up many of Hume's views on rationalism, without his skepticism. They used logic to construct unified objects out of Hume's impressions and ideas. Nelson Goodman, whose views grew out of those of the logical positivists, dealt with induction by appealing to "entrenchment," that is, we project those qualities into the future which have been traditionally accepted, a highly conservative view (Fuller 2000, pp. 87–88; 2014d, p. 141).

Fuller sees Darwin as an inheritor of Epicurus' atomism and randomness. Darwin is also an inheritor of Hume's skepticism concerning and criticism of divine design. That is not to say that Darwin was a very diligent reader of Hume, let alone having studied Epicurus. Rather Darwin was a developer of ideas of theirs that had come down through history. "Darwin's bulldog," T. H. Huxley, was responsible for the revival of Hume's ideas (Fuller 2008, p. 114). Huxley also invented the terms "agnosticism" for not knowing whether God exists, and "epiphenomenalism," for denial that the mind has physical influence.

There is an important moral motivation for Fuller's negative judgment on Epicurus and Hume. He contrasts the therapeutic approach of Epicurus (2011a, p. 261) with his own proactive approach. He also associates Darwin with the "passivist tradition [which] includes Malthus, Spencer and Darwin" (Fuller 2007a, p. 34). According to Fuller the Darwin-Epicurus tradition emphasizes the avoidance of suffering rather than seeing suffering as means to advancement. "No Pain no gain" (Fuller 2015). This is linked to Fuller's support of pursuing Humanity 2.0, opposing the precautionary approach of ecology. Fuller advocates risk taking, such as self-experimentation and the undertaking of possibly dangerous technological projects.

Religion as Historical Motivation for Science

The most controversial activity of Fuller was his testimony on behalf of the defendants in *Tammy Kitzmiller, et al. v. Dover Area School District, et al.* (2005), in which Fuller took the side of the supporters of Intelligent Design. Fuller testified as a philosopher, historian, and sociologist of science. Fuller saw this action consistent with his belief that intellectuals in general, and science studies people in particular, should enter the public arena (Fuller 2008, p. 11). This participation greatly upset many members of the science studies community. The testimony gave ammunition to opponents of science studies from the "Science Wars" debate. To the "Science Warriors" opposing social studies of science, the most vituperative of whom was Norman Levitt, this was claimed to be clear evidence of what they are searching for, an alliance of postmodernist humanists, social scientists, and science studies people with the Intelligent Design movement. This was despite the fact that Fuller is not a postmodernist and supports several features of positivism. Those who wished to discredit science studies had long claimed parallels between religious fundamentalist opponents of mainstream science and the supposedly leftist science studies community.

In his work *Science* versus *Religion?* (2007b) and *Dissent on Descent* (2008), Fuller presents a sweeping historical and philosophical account of the notion of divine design, showing its importance in the origins of modern science. This part of Fuller's account is indisputable. (See discussion of Grayling below.) The much more controversial claim of Fuller in the Kitzmiller testimony and in his written works is that all that is best in contemporary science involves a design-based stance and that Darwinism erodes the motivations of science itself.

In defending his version of genuine (design oriented) science, Fuller suggests that Darwin may not be a true scientist (Fuller 2008, p. 40). In contrast, Fuller claims, Intelligent Design theory is in the mainstream of Western science, leading through Mendel's genetics, biotechnology, artificial life, and nanotechnology to modern science, which Fuller claims has no need of Darwinism. Fuller wishes to contrast these high-tech fields with natural history, which he associates with Darwin and disparages in its contemporary form in evolutionary biology.

Fuller's brief is a more sophisticated and historically informed version of the creationist, and ID claims that Darwin's world view makes the world devoid of meaning. This is a view that, indeed, some but by no means all neo-Darwinists hold. The historian of population genetics William Provine held this, as does the physicist Steve Weinberg, with respect to the cosmos (Weinberg 2007).

Fuller accurately counters claims that irreligion was the driving force of early modern science. For instance, A. C. Grayling in a negative review of Fuller asserts that the rise of early modern science was prompted by atheism (Grayling 2008). The historical part of Fuller's argument is very accurate concerning early modern science. Figures such as Boyle, Newton, Leibniz, and many other figures of seventeenth-century science saw their religion as tied with their science. Further, in the eighteenth- and nineteenth-century science, biologists the classifier of species, Linnaeus, as well as the eighteenth-century British naturalist churchmen assumed nature was divinely designed. Going beyond the usual accounts of the role of theology for early modern scientists, Fuller includes the founder of genetics, Gregor Mendel, who was a monk. Fuller further makes an ingenious but tenuous educational tie between Mendel and the chemist Joseph Priestley (Fuller 2008, pp. 83–84).

Fuller could claim that religious motivation need not undermine the validity of the theory produced. However, he goes much further and sees religious inspiration as constitutive of modern science.

Unfortunately, Judge John E. Jones III, who admirably paid strict attention to the testimony on both sides, largely misconstrued the point of this major part of Fuller's presentation. The judge accepted that ID was theology based, but despite Fuller's account of the design basis of the excellent science of Boyle, Newton et al. concluded that ID could not be genuine science if religiously motivated.

Fuller fills out his very well informed and richly informed historical account with delineation of varieties of theodicy. He discusses Leibniz as well as the conflict between the approaches to the perfection of nature of Nicolas Malebranche and Pierre Gassendi. Fuller claims that Leibniz's concern in theodicy was with the moral education of humankind, using St. Augustine's claim that light needs shadows to justify the existence of evil. Gassendi, the Christian atomist, claimed that God's creations are perfect but that our intellect is incapable of fully discerning the nature of the perfections. Father Nicolas Malebranche in contrast claimed that the imperfections of created things were part of a trade-off to produce the greatest total overall perfection consistent with simplicity and uniformity (Fuller 2008, pp. 183–186). Fuller aptly notes that Roman Catholic ID advocate Michael Behe and Catholic theistic evolutionist Kenneth Miller, who are on opposite sides in Kitzmiller, replay the Malebranche vs. Gassendi debate.

Fuller represents an analogy between supernatural causes and theoretical or unobservable terms (Fuller 2008, p. 115; 2010, pp. 103–104; 2011a, p. 179). Fuller also writes on "supernatural induction" (2008, p. 115) in which he claims that the hypothesis of God as a designer is like the hypotheses of the existence genes and atoms in the nineteenth century, as both God and theoretical entities are unobservable. True, both theoretical entities and God are unobservable, but there the similarity ends. Atoms and genes were given laws of combination and/or motion. Presumably humans cannot give detailed accounts of God's purposes and plans, but Fuller advocates an ID that would do so.

William T. Lynch (2016) gives a well-informed and documented case for an evolutionary psychology account of science to counter Fuller's claim that belief in evolution takes away the motivation to do science. He appeals to competition and social selection to account for the progress of science rather in the manner of David Hull's *Science as Process* (1988). One danger of this might be the over reliance of selection mechanisms similar to the pan-selectionism of the evolutionary psychologists. It is often argued that the ability to do mathematical science has no survival value to be selected for in early human evolution. However, the capacities for the sort of formal thought involved in science are immanent in basic human capacities for spatial imagery and counting, which, indefinitely extended and reiterated, can yield the sort of advanced mathematical structures used in science. These capacities are spandrels. Often the reason people think that advanced abstract math (and its applications) are not evolved is that they do not recognize to what extent abstract math results from primitive visualization (Formalism, logicisim, and the Bourbaki-based "new math" misleadingly downplayed this role).

Darwin Not a Scientist?

The less justified flip side of Fuller's account of the history of the role of the design hypothesis is his downgrading of Darwin. Chapter 2 of *Dissent on Descent* (2008) is titled "Was Darwin really a scientist?" Of course, Darwin was not a scientist in our contemporary, professional sense. "Scientist" was first used by William Whewell in reply to Samuel Taylor Coleridge at the British Association of Science in 1832 (Snyder 2011, pp. 1–2). It was after the *Origin of Species* that Darwin's "bulldog," T. H. Huxley, forged professional science and the professional role of the scientist in Britain. Huxley had a lower middle class background, while Darwin was wealthy. Darwin never had to work, inheriting wealth both from his parents and from his wife. He was an amateur if one counts only those as professionals who work in an institute of higher education or research. However, Darwin was a professional insofar as his work was published and respected by the members of scientific societies of London. On institutional grounds one could question many other notable early nineteenth-century students of nature were scientists in the modern sense.

Fuller strongly contrasts Darwin's natural history approach with the developing laboratory-based biology of the time. "Darwin and even Gould felt more comfortable with the title 'natural historian' than 'natural scientist'" (Fuller 2007b, p. 84). However, Darwin did simple experiments that did modify or create conditions. In his work on distribution of species, he tested survival of seeds and hibernating snails to survive. He also did experiments with slave-making ants. In his botanical work, Darwin did many experimental manipulations. His conclusions concerning the source of root growth and of upward stem growth were vindicated many decades later. He also did experiments on the triggering mechanism of the Venus flytrap (Chamovitz 2013, pp. 14–16, 68, 118–119). In his last book, on earthworms, he tested worms' senses and behavior with methods now recognized as ahead of his time (Darwin 2015 [1882]; Gould 1983, p. 129).

Fuller similarly contrasts recent scientists involved with evolution and population genetics as backward compared to laboratory scientists, computer modelers, and ID advocates. Fuller claims Stephen Jay Gould represents the (for Fuller passé) natural history sources of evolutionary theory. Fuller claims that Gould has "general ignorance and disdain for laboratory based science" (2008, p. 139). This is hardly accurate, insofar as many of the latter's *Natural History* articles describe and praise various laboratory studies. Gould, dismissed as an old school natural historian by Fuller, in contrast to the high-tech ID advocates, actually published two studies based on computer modeling early in his career (Raup and Gould 1974; Gould et al. 1977). Fuller calls both Darwin and Gould "masters of the mollusk" (2006c, p. 135), which, despite the attractive alliteration, is false for Darwin. Darwin studied barnacles, which are crustaceans, like lobsters and crabs, rather than mollusks, like Gould's snails. Fuller also classifies Richard Lewontin as a population geneticist in contrast to lab-based molecular biologists (Fuller 2007a). Actually, Lewontin pioneered the lab technique of gel electrophoresis of proteins as a means of measuring genetic variation.

Fuller does not try, as some have done, to downgrade Darwin's achievement based on Alfred Russel Wallace's independent discovery of natural selection. However, perhaps it is because Wallace's spiritualism and special place for the human mind as not naturally selected, inconsistent with the meaningless of Darwinian universe thesis, that leads to Fuller's omission of Wallace.

Chance and Darwin

Fuller also contrasts the design tradition with Darwin's approach, calling the latter "chance based" (Fuller 2014a, pp. 87–88). Darwin's theory is claimed to be founded on random variations that are chaotic in a meaningless universe. However, Darwin's variations are "random" only in the sense of not directed toward improved fitness or function. Darwin thought variations to be deterministic but with microscopic mechanisms unknown to us. Most working biologists today also understand Darwinism as having an underlying strict determinism. Also, the claim that natural selection is "chance based" ignores that selection as opposed to mutation is not chance based. Darwin's reference to chance is less extreme than that of those who present evolution as purely random and then conclude its probability is miniscule (cosmologist Fred Hoyle and most creationists and ID advocates). However, Fuller's characterization is a misrepresentation of Darwin and neo-Darwinism, but may apply to so-called non-Darwinian or neutral evolution of the Japanese school. One controversial interpretation of natural selection is that drift is "biology's first law," (McShea and Brandon) where drift, as the zero-force law of biology, is analogous in biology to Newton's first law in physics. That is, deviations from randomness are accounted for by evolutionary forces such as selection just as deviations from uniform straight line motion are accounted for by physical forces in Newtonian physics. Fuller treats Hardy-Weinberg equilibrium, that is, the simple genetic result of lack of selection, mutation, and migration, as making results of natural selection merely "a variety of exceptions" to equilibrium (Fuller 2008, p. 117). One could just as well say that motion under pure gravitational force without electromagnetic attraction involves "a variety of exceptions" to Newton's first law, which some, such as Henri Poincaré and Erwin Schrödinger, treat as a tautological definition. Also, both Newton's first law as and Mendelian Hardy-Weinberg equilibrium as well as genetic drift can be understood as "ideals of natural order" in Stephen Toulmin's coinage involving a variety of idealizations in which the exceptions or deviations can be explained in terms of forces (Toulmin 1963).

Fuller constructs a dilemma. On the one hand, there is the purely selectionist version of evolution of Dawkins and others who neglect non-selection factors. This is a convenient opposing view, for ID supporters usually present the only alternative to pure selectionism to be ID. If some feature cannot be accounted for by selection, it must be due to Divine action. On the other hand, there are broader conceptions of evolution in contemporary theory. These include embryonic factors (evo-devo), purely physical form and constraints, and internal physical processes. Fuller calls the broader conception "inordinately equivocal" (2008, p. 110). However, the broader conception of evolution hardly makes it not evolution. By the way, Darwin himself incorporated purposive neo-Lamarckism in later editions of the *Origin*. Unlike the co-discoverer of natural selection, Alfred Wallace, Darwin was not a pure selectionist. Darwin's book might have been called *The Origin of Species by Natural Selection and Many of Other Things.*

As a minor historical point, Fuller often attributes to Lamarck what is really neo-Lamarckism (2008, p. 98) emphasizing "conscious striving" (2011a, p. 19) and "deliberate effort" (2011a, p. 148) although elsewhere he writes of Lamarck's "orthogenetic view" (2006a, p. 171) and of "matter's self-directing tendencies" (2010, p. 100). The historical point of Lamarck's main theory was that of an inner drive that led inorganic matter through the various, increasingly complex life forms. This is closer to what was called "orthogenesis" than to what later was called Lamarckianism. Every chunk of earthly matter traveled this trajectory. It is just that some started earlier than others, leaving humans with a very early start contemporaneous with invertebrates that had a much later start evolving. The inheritance of acquired characteristic in Lamarck accounts for the side branches from the main highway of evolution, such as insects and birds. Later theories that were solely based on inheritance of acquired characteristics by exercise or learning are now called "neo-Lamarckian." If one means by Lamarckianism the inheritance of characteristics produced by exercise or mental effort, Lamarck and the later Darwin were "Lamarckians" to the same degree. Lamarck was mostly orthogeneticist and only a fraction "Lamarckian," while Darwin was mostly a selectionist and a fraction "Lamarckian." In fairness to Fuller, every biology textbook and most books and articles by biologists and others make this identification of the historical Lamarck with purposive neo-Lamarckism.

With historical accuracy Fuller strengthens his case concerning the theological roots of parts of probability by noting the theological background of theorem of Reverend Bayes and others of the period (Fuller 2014c, pp. 79–80). Laws of probability limit sheer randomness in processes. If probability tames and limits chance, should not this apply to statistical processes and trends in biology?

Although Fuller sees the probabilistic aspect of Darwinism as a concession to Epicurus and Hume, Fuller's counter-tradition of twentieth-century technological design has significant elements of probability theory as well. Norbert Wiener's cybernetics and Claude Shannon's information theory involve probability and randomness. Wiener's analysis of random walk-like "Wiener processes" and smoothing of time series involve randomness and probability. Shannon's definition of information incorporates probabilistic information, parallel to entropy in statistical mechanics. Genetic programming of computers uses variation and natural selection.

Despite his holding of ultimate determinism, one of Darwin's greatest innovations was to suggest that statistical thinking as the basis of a science. Darwin's innovative introduction of statistical thinking in biology was stimulated by his reading of a summary of Adolphe Quetelet's early statistical sociology. Darwin's use of statistical thinking may have led the physicist Ludwig Boltzmann to introduce probability as the basis of his statistical mechanics. Isaac Asimov once suggested, but was too modest to publish, the speculation that Boltzmann's introduction of probability into statistical mechanics was inspired by his great appreciation of Darwin (Seeger and Cohen, Cohen). Boltzmann's statistical mechanics, despite his belief in a determinist, Newtonian substratum, may have contributed to the development of indeterminacy. (Charles S. Peirce, an anti-Darwinian evolutionist, was one of several later nineteenth-century figures that advocated real indeterminism and recognized the central significance of the introduction of statistical thinking by Darwin.) Quantum field theorist Sylvain Schweber became an historian of Darwinism by tracing James Clerk Maxwell's own version of the introduction of probability into statistical mechanics to the same Scottish political economy and anthropology absorbed by Darwin (Schweber 1982). Would Fuller reject statistical mechanics in physics? Consistently, Fuller rejects the Scottish tradition of political economy and anthropology insofar as Hume and Adam Smith were influences on Darwin's thought.

On Fuller's account it would seem to follow that quantum mechanics with its genuine indeterminism would not count as genuine science. It seems Einstein's claim that God does not play dice is for Fuller less metaphorical than Einstein took it. It is true that both neo-Darwinism and quantum theory, in different ways, have led philosophers of science to modify Hempel's deductive model of science. Chaos theory, with its deterministic underpinnings but with empirical unpredictability, has some resemblance to Darwin's own views in that, although ultimate determinism is assumed, predictability is impossible, because knowledge of the literally infinitesimal perturbations that make for significant differences in trajectories is beneath our ability to measure, even in principle. Would this make quantum mechanics, chaos theory, and statistical mechanics not real sciences? He does not seem to think that chaos "theory" (in scare quotes), for instance, is not a real theory (Fuller 2008, p. 87).

The Engineering and Information Tradition in Biology

Fuller emphasizes that Erwin Schrödinger was a follower of Boltzmann (2008, pp. 207–210). Schrödinger, like Einstein, held out for a (non-forthcoming) deterministic version of subatomic physics, despite his teacher Franz S. Exner being an early indeterminist. Schrödinger also questioned the existence of quantum jumps. Schrödinger's *What Is Life?* (1944) was an inspiration to Francis Crick, Maurice Wilkins, and others who were disillusioned with the outcome of physics in the atom bomb and turned to molecular genetics, a field they erroneously thought would lack the ethical dilemmas of the former. Fuller includes Schrödinger as a major figure in his history of the design tradition. However, central to Schrödinger's book is the notion of negentropy as characteristic of life. Entropy is a statistical concept. Its increase is statistical rather than wholly uniform, as in earlier classical thermodynamics before statistical mechanics. Thus, even Schrödinger holds that a statistical quantity is central to biology. Significantly, Schrödinger, like his ideal physicist Boltzmann, was a Darwinist.

Fuller also links ID's design-based approach with contemporary information science. He refers to Charles Babbage's *Bridgewater Treatise* where divine creation is described as analogous to the writing of a computer program that activated the creation of species at successive times (Fuller 2007b, pp. 85–89; 2008, p. 189). Fuller links the Unitarian religious training or alliances of figures in cybernetics and artificial intelligence such as Norbert Wiener and Herbert Simon to the Intelligent Design approach (2014a, p. 48). One might add George Boole, inventor of the Boolean algebra used in computing, who was sympathetic to Unitarianism as well as to Judaism (McHale 1985, p. 185, 195, 200).

For Fuller, the rise of disciplines such as artificial intelligence and artificial life, as well as the general computer orientation of bioinformatics and contemporary molecular biology, is cutting edge science and is a kind of "Intelligent Design" theory. In the sense that an intelligent human designed them, they are. Fuller wishes to extrapolate from this via his Scotistic continuum to God as designer.

By linking ID theory, generally castigated as pseudo-science by the biology community, or, at best, as "dead science" by Philip Kitcher (2007, p. 8, 12) with the grand tradition of design in nature of the Greats of early modern science, as well as with cutting edge biotechnology and molecular biology, Fuller has an ingenious rhetorical strategy to validate the credentials of Intelligent Design.

ID as Research Program

The only problem is that, despite Fuller's allusions to ID as an ongoing research program, there seems to be no *positive* research going on among ID advocates. All their claims with respect to the origin of species are negative claims. Michael Behe revives a molecular version of the troublesome objection by the tragic, Catholic evolutionist St. George Jackson Mivart that partially evolved organs would not be of selective value and that therefore the intermediate steps could not survive. Dembski argues for the irreducible information content of living things, impossible to achieve by natural selection. Other ID supporters deny that the Cambrian explosion could have arisen by Darwinian selection (borrowing in part from Gould). However, none of the design advocates use their appeal to the mysterious Intelligent Designer to articulate explanations in detail or to make explanations, predictions, or retrodictions of the existence of particular organisms or features of organisms. Fuller himself, to his credit, advocates making more specific explanations in terms of God's purposes, rather than simply saying God did it. However, in actually existing ID, this is very much a promissory note, as yet uncashed.

Joseph Agassi once suggested that a science is conceivable that functioned solely in terms of theories and refutations without any positive developments of the theory. Only such an extreme and hypothetical "left Popperian" approach fits the purely negative claims of contemporary ID. (Agassi himself is an opponent of ID and of theodicy.)

To counter the fact that actually existing ID makes no new predictions or explanations of particular phenomena, Fuller claims that the use of new predictions or explanations as a criterion for superiority of a scientific theory to another is at best naive (Fuller 2011c, pp. 567–568; 2014c, pp. 77–78). This is a rejection of Popper, whom Fuller admires (2004). He notes, as do Popper and most contemporary philosophers of science, that evidence is relative to a theoretical background and rejects the notion of evidence as a body of recorded descriptions. However, even with respect to theory-laden evidence, ID theorists don't offer any. Certain elegant theories, such as Hermann Weyl's version of a unification of gravitation and electricity and David Bohm's quantum theory, were rejected because they gave no new explanations or predictions in comparison with previously existing theories. (In Bohm's case this was in part due to his Marxism.)

Fuller claims that since ID is a young science, it cannot be expected to yield predictions and descriptions that are not accounted for by natural selection. He advocates "affirmative action" for ID. Yet there are many very preliminary and rudimentary theories without enough significant tests that are presently overwhelmed by well-confirmed sciences that dominate their field. The philosopher of science, Paul Feyerabend, wished alternative theories to proliferate as much as possible, but affirmative action for all these alternative theories or viewpoints that purport to be theories could be costly. Why should ID get affirmative action and not holistic medicine or homeopathy? Fuller's theory of knowledge management would evaluate each proposed research topic in terms of the expense and benefits to the public.

Creation from Nothing

Immanuel Mesthene (1967) in an early STS defense of a kind of moderate technocracy nicely summarizes the claim that modern science completely denies the existence of the surd in nature. However, other cultural forms of knowledge of nature did yield contributions to description of nature that are not modern Western science, but which denied the ultimate comprehensibility of nature at the micro-level. Ancient Greek examples are Aristotelian matter and the Platonic matrix or *chora*. Werner Heisenberg later appealed to retrospectively justify his indeterminacy, which has been claimed to have come from his adolescent reading of the *Timaeus*. Chinese science, with its notion of ultimate indeterminacy or vagueness of measurement (Sivin 1995), is a prime example of this view. At the end of his book *Science*, (1997) Fuller claims rightly that contemporary science is moving in the direction of the governmentally directed and purely practically oriented science found in the ancient Asiatic despotisms, and is losing the commitment to absolute truth of early modern science. Fuller might say the turn to quantum mechanics, indeterminacy, and chaos is simply another sign of this degeneration in postmodern and neo-liberal science and society. Chaos theory is close to postmodern architecture in reviving classical styles, but adding them to new notions.

A feature of the Judeo-Christian notion of divine creation that Fuller, rather surprisingly, does not use is the creation of the universe from nothing, *creatio ex nihilo*. Fuller usually describes divine creation in ways consistent with the incomprehensible surd aspect of matter. Fuller discusses God's design as being limited by the recalcitrance of matter (2008, pp. 172–173). This seems to reject the *creatio ex nihilo* view in traditional Judaism and Christianity since the first or second century. Elsewhere, inconsistently, Fuller seems to accept *creatio ex nihilo* (Fuller 2014c, p. 20, 22). What distinguishes Christianity from pagan religion both East and West (Greek polytheism as well as Hinduism and Buddhism), at least since Alexandrian neo-Platonist Philo Judaeus, is that God created *everything* (not only form but matter) from nothingness. In other creation myths, God or gods simply mold pre-existing stuff (matter) that is not itself structured. Possibly the "without form and void" and "darkness over the face of the deep" in the opening lines of *Genesis* were originally meant this way, but since the early Common Era have been taken as literally nothingness. The God of the *Genesis*, at least as interpreted for some two millennia, if not three, is omnipotent, omniscient, and rational. God is not restricted by matter in creation. This view fits well with the view of pre-twentieth-century Western science as rejecting any inexplicable aspect of nature. Hans Jonas (1967) claimed to derive both rationalist and empiricist theory of knowledge as alternative responses to the claim that through creation from nothing the universe is thoroughly rational. The two differed, according to Jonas, on the extent of human access to God's rationality.

Theistic Evolution

Fuller emphasizes that Newton's use of divine plan in understanding the universe is valuable, but Newton's occasional intervention of to realign the planets in their orbits is not. (Leibniz accused Newton's God of being a lazy God for not perfectly pre-programming the universe.) However, the standard ID account of individual organisms seems to accept both sides of

Newton. ID seems to more resemble the individual acts of intervention by Divine power in speciation, rather than the laying down of an overall plan or laws or the universe. Species creation in ID's account seems like a mini version of Cuvier's catastrophism.

Fuller sees Charles Babbage's so-called Ninth Bridgewater Treatise on computer programming as a precursor of later artificial life programs (Fuller 2008, pp. 185–189; 2009, p. 189). However, if God were to develop programs that evolved species one from another, rather than successively creating them, this version of creationism could be turned into theistic evolution.

Fuller rejects theistic evolution for not having God make a difference. Theistic evolutionists' "religious beliefs spin as decorative but cognitively as metaphysical idle wheels," (Fuller 2008, p. 109) criticizing the single act, all at once, model of creation of organisms, in such a way as to make them instantly intelligible. There is a whole continuum of degrees of intervention between fiat creation and a theistic evolution that totally detaches the naturalist process from Divine oversight. In this version the God of theistic evolution becomes like the detached gods of Epicurus. At one extreme, Darwin's first American advocate, Asa Gray, left natural selection alone, but claimed God was guiding the apparently random variations. Michael Behe's more recent formulation of ID moves closer to theistic evolution by granting that much biological change is driven by standard mutations and natural selection, but that certain simultaneous sets of mutations are due to Intelligent Design (Behe 2007). Some theistic evolutionists, such as Fuller's beloved Teilhard de Chardin, do modify natural selection by claiming that there is an overall purpose to evolution. Theistic evolution could treat God as a tinkering designer rather than a creator of whole species by fiat. Intelligent Design might move closer to theistic evolution if the acts of tinkering were small enough. It is not clear in ID the status of the successive creations.

Control of Nature

There is a strong theme of human control and planning of nature in Western history. Humans as completers of God's work or stewards of nature would certainly fits with and may be the source of this conception. Nevertheless, one may ask, does this legitimate ID in the specific sense in accounting for the past? Lurking less obviously in the back of Fuller's model of creationism with humans made in the image of God is Fuller's belief that the duty and destiny of humans is to control nature. This is made more explicit and much further elaborated in *Humanity 2.0*. While Fuller supports Karl Popper's critical rationalism against the defense of dogmatism by Kuhn and Michael Polanyi, Fuller apparently rejects Popper's strictures against global planning, either of society or the universe.

In the appeal to design in biotechnology and computers, there is an apparent or hinted justification of belief in ID as an explanation of biological species. Fuller is probably right that (at least on an ideological level) notions of humans engineering and designing organisms will dominate the new biology. There is, however, the issue of how much of the bioengineering will be done from scratch, as opposed to making minor modifications of already existing organisms or molecules synthesized by microorganisms. Even the restriction enzymes basic to gene splicing were natural enzymes discovered in bacteria, not constructions by synthesis. Similarly, the CRISPR apparatus, now used for genetic editing, was found naturally occurring in microorganisms, not thought up *ab novo* and synthesized in the lab by a biotechnologist.

Fuller raises doubts about the status of evolution as a science by noting that the Nobel Prize has never been awarded for work in evolution as it has been in chemistry and physics (2008, p. 204). Nobel also has not given a Nobel Prize for mathematics, because he thought it was not directly practical. (The story that Nobel omitted a prize in mathematics because his spouse had an affair with a mathematician is apocryphal.) Einstein did not receive a prize for relativity theory, but for his work on the photoelectric effect, which was clearly seen as more relevant to practical applications at the time. Einstein's Special Relativity paper was turned over for evaluation of quality by the Nobel committee to a medical occultist, who found it lacking. The Nobel Prize for physiology and medicine is the closest to a prize for biology. Work on evolutionary theory was neither physiology nor in medicine though there have been recent attempts at Darwinian medicine. It is understandable that no Nobel Prizes have been given for work in evolution but not on the basis that it is not science.

An example of skepticism about Darwinism that Fuller might have better appealed to is that of the French medical establishment. Perhaps because of the heritage of Lamarck and Cuvier Darwin is less highly regarded in France than in Britain or Germany. Darwin's reception in France was not helped by the fact that the first French translator was a woman and an anarchist. At least as late as the 1970s, many professors of biology and medicine in France dismissed Darwinism as mere speculation (Gayon).

Newton and Darwin

Fuller also claims that Darwin's causal analysis is "feeble" and that Darwin is "no Newton" (2008, p. 205). This ignores the innovative and original nature of natural selection as a causal mechanism. In his *Humanity 2.0* Fuller is certainly correct to criticize the role of solely "humble" underlaborer publically played by many contemporary philosophers of science. He particularly takes umbrage at the Darwinian epigones in the philosophy of biology (2011b, pp. 164–165).

However, Fuller overstates the unity and historical continuity of Newtonianism as opposed to Darwinism. He notes that we speak of "neo-Darwinism" but not "neo-Newtonianism." Fuller thinks that while Darwinism had to add Mendelian genetics to flourish in the twentieth century, Newtonianism developed continually and successfully based solely on its axioms. However, Fuller's nemesis, Thomas Kuhn, was forced to add a footnote to the second edition of *Structure* in 1970 noting that he had overestimated the unity of Newtonianism in the eighteenth century, referring to Clifford Truesdell's (1960) work on Euler and the Bernoullis (1970, p. 31). In the original edition of *Structure*, Kuhn noted that "Newtonianism" in chemistry and optics was quite different from the austere Newtonianism of rational and celestial mechanics. However, even within rational mechanics, the development of "Newtonian" mechanics involved many additions and modifications. Jacob Bernoulli gave the formulation the general differential equation version of Newton's second laws used ever since. Leonhard Euler developed Newtonian mechanics further to apply to rigid and flexible bodies. Further innovations were involved in applying Newtonian mechanics to fluids and hydrodynamics (Dusek 1966a).

One can go further than Kuhn and Truesdell in claiming that Lagrangian and Hamiltonian mechanics are quite different in formulation from Newton's original mechanics as are relativity and quantum mechanics (Dusek 1966b). Least action principles from Leibniz through William Rowan Hamilton, Max Planck, and David Hilbert have a teleological aspect absent from the past-to-future efficient causality of the differential formulation of Newton and John Bernoulli. Hamiltonian mechanics, but not Newton's Laws, is necessary to formulate quantum mechanics, as is

Hamilton's light-ray particle analogy. The so-called potential theory formulation of classical mechanics yields yet another conceptually different science, yielding, in the end, all of Aristotle's four causes in nineteenth-century mechanics (Dusek 2000).

Fuller blames Darwinism for the recent philosophy of science rejection of the hypothetico-deductive unified approach for one where "Theory is no more than a family of models" (Fuller 2011a, p. 170). But later, nineteenth century, classical mechanics can be formulated in the model version. Also, if the family of models is based on logical model theory, there are connections.

According to Fuller, physics has a unified deductive structure while evolution does not. He sees the newer model-based accounts of science as a degeneration from the hypothetico-deductive method inspired by need to account for explanation in evolution. However, accounts of physics such as those of Nancy Cartwright (1983) and others suggest that even physics lacks the straightforward deductive prediction structure Fuller claims is characteristic of genuine science. Nor only Darwinian evolution but quantum mechanics has led philosophers of science to present alternatives to Hempel's deductive model of explanation. The indeterminism of quantum theory undermines Carl Hempel's account of statistical explanation. Bas Van Fraassen's model approach to theory-based account of theories and pragmatic theory of explanation is very much tied up with his work on quantum mechanics (1991). However, his student Elisabeth Lloyd (1994) applied it to evolutionary theory.

Fuller is right in his claim that much biology can go on as usual without evolution or Darwinism. He notes that only 10% of biology articles explicitly refer to evolution. This may be because Darwinism plays a unifying role in biology analogous to particle physics in the rest of physics and chemistry. Particle physics articles likewise account for a small percentage of the total number of physics articles. Working chemical physicists and solid-state physicists do not refer to the Standard Model in elementary particle theory, as quarks play no role in their accounts of atoms or metals. Similarly, this is the case for chemical planetary physics, celestial mechanics, and many other branches of physics. The French physicist Jean-Marc Lévy-Leblond suggested post-war particle physics mostly have not had applications, as did the earlier quantum mechanics, to nuclear weapons and power plants, but gives practitioners assurance that physical science has a basis in ultimate entities (1975, pp. 136–175). Darwinism plays this role for the unification for biology but in a less directly reductionist and more historical manner.

Fuller is an advocate of unity of science. (See Chap. 2.) A further counter to Fuller's defense of molecular biology's reductionism might be that anti-reductionism with so-called emergence of higher level organization explained by different principles than the lower level affords a basis for a distinction between humans and other non-linguistic animals. If mind and society are higher levels of organization from molecules and organisms, then a distinction between humans and other animals can be claimed without appealing to humans being in the image of God. As we have seen, Fuller tends to lump anti-reductionist biologists, such as Gould and Lewontin with those who reduce humans to animals, which the Gould does not do. Fuller also pours cold water on Chomsky's differentiation of humans by possession of recursive grammar to support the "(allegedly) unique species capacity for language." Elimination of emergence and of linguistic innateness leaves the differentiation of humans from animals solely based on humans being the image of God.

Oddly, Fuller defends sociobiology (an account of society and culture based on evolutionary biology) and sees its critics who defend human action as purposive and the human mind as having a level of organization about that of biology as merely politically correct, and Gnostic based on Ullica Segerstrale's (2000) biased, pro-sociobiology account, which he treats as fair and balanced. Gould sees human social evolution as neo-Lamarckian, not Darwinian, and even alleged genetic determinist Richard Dawkins distinguishes cultural memes from biological genes. Also, Fuller defends the reality of human races. (See Chap. 6 on Gnosticism.)

Rifkin's *Algeny*: Similarities to Fuller

There are several parallels of Fuller's case for ID to Jeremy Rifkin's *Algeny* (1983). Rifkin, although less scholarly than Fuller, sets Darwinism and biotechnology in a broad social setting. Rifkin claims that Darwinism reflects industrial society and biotechnology reflects post-industrial society. (Rifkin mentions competition, division of labor, immigration, and other terms borrowed in Darwin's work.) Rifkin gives a mostly accurate, readable popularization of the work of Robert Young, Sylvain Schweber, and others, on the industrial and capitalist terminology in Darwin's work (though Rifkin emphasizes French positivist "industrialism" rather than the Marx's "capitalism"). Rifkin, like Fuller, claimed that the informational and computer orientation of biotechnology (more thoroughgoing today with bioinformatics than when he wrote) makes biotech the expression of post-industrial or information society. A difference of Rifkin's thesis is that model of nature and biotech with post-industrial society he views the rise of biotech will be tied to the young earth creationism of Duane Gish and Henry Morris. Rifkin is on to something in the first part his book, but spoils it with very naive young earth creationism of later chapters. Fuller, on his part, links molecular biology, biotechnology, and artificial life to ID, not to young earth creationism.

Fuller ties modern design-oriented computer-enhanced bioscience to the role of humans "playing God." This phase is unfortunately and unreflectively overused in the popular debates over genetic engineering. It is a theme that Fuller takes much more seriously than the usual superficial, metaphorical use of it and develops favorably philosophically in his *Humanity 2.0*.

Fuller certainly is correct in this claim. Indeed, he could make an even stronger and more well-documented case for this than he does. He cites (2008, p. 200) one passage from Norbert Wiener's *God and Golem, Inc.* (1964) on the power of the devil, but there are numerous discussions throughout that work dealing with the theme that the engineer is to his or her device as God is to creatures.

Parallel of Methodological Debates in Sociology and in Evolutionary Theory

Finally, using classical sociological theory disputes, Fuller criticizes contemporary Darwinian evolutionary theory for mixing or shifting between the lawful (nomological) and the individual-descriptive (idiographic). He correctly notes that methodologists of the social sciences are well placed to analyze these (purported) confusions in evolutionary theory. He is correct that the Modern Synthesis of neo-Darwinism, as well as explicitly moves of the competing version of Darwinian evolution of Gould, shifts between or mixes the two approaches. However, this is hardly a criticism of Darwinian evolution. Heinrich Rickert, the neo-Kantian analyst of science and history granted that there were fields, such as geology, where the two approaches were mixed. Max Weber's methodology of the social science involves combines explanation (*erklaerung*) with understanding (*verstehen*). Evolutionary theory similarly combines idiographic narration with nomological model building. We think this is what the synthetic theory of evolution and many of its modifications and descendants (such as punctuated equilibrium and evo-devo) do. Perhaps significant for Fuller's dismissal of the natural history approach is that his rejection of verstehen, hermeneutics, and the version of historicism rejected for an account of science closer to that of the logical positivists. Oddly, Fuller for all his scholarship on human history, rejects, or, at least, denigrates, the historical aspect of the study of nature.

Mathematical ecology is totally neglected in Fuller's account. Fuller is skeptical of and suspicious of ecology because of its political applications, in that it calls for restrictions on certain power technologies and bemoaning current mass extinction. Fuller might well accept Lynn White's claim that Christianity is the source of extreme exploitation of nature, but would see this as to Christianity's credit rather than deficit, as does White (1967). This may be one reason for Fuller's neglecting of scientific ecology.

Mathematical ecology is not the same as older, purely commonsensical, relatively atheoretical natural history. In the hands of such theorists as its major founder, G. Evelyn Hutchinson, his students, most notably Robert MacArthur, and other practitioners such as Robert May and Joan Roughgarden, theoretical ecology is as mathematically sophisticated as chemistry and physics. This mathematical, theoretical ecology blurs the line between laboratory science, computer modeling, and the older natural history. There are field observations that are not totally uncontrolled. An example is E. O. Wilson and Daniel Simberloff's covering with plastic, Christo-like small Florida islands, using insecticide to eliminate the insect inhabitants, and then tracking the patterns of repopulation. Paul Ehrlich at Stanford tracked the extinction of the checkered white butterfly (more to the horror of some students and colleagues who wished to artificially save the population). This may be natural history observation but was not mere casual, common sense observation.

It is true that many working evolutionists portray their methodology and explanatory strategy as wholly nomological, aping the physicists, as do the many social scientists that suffer from so-called physics envy. However, this methodological self-misunderstanding does not vitiate the effectiveness of the actual mixed method applied. Some evolutionists, like Gould, have explicitly emphasized, indeed reveled in, the historical aspect of geology and paleontology, and hence in the historical aspect of macro-evolutionary description of the paths of evolution. This does not eliminate nomological treatments of the material.

Fuller's discussions of ID are packed with fascinating historical background and original analogies between religious and scientific positions and debates, and ingenious rhetorical strategies. More questionable is his positioning of ID as scientific, while claiming that evolutionary theory and research are not scientific.

References

Behe, M. (2007). *The Edge of Evolution: The Search for the Limits of Darwinism.* Glencoe: The Free Press.

Bonaventura, S. (1952). *The Mind's Road to God.* Indianapolis: Liberal Arts Press.

Cartwright, N. (1983). *How the Laws of Physics Lie.* Oxford: Oxford University Press.

Chamovitz, D. (2013). *What a Plant Knows: A Guide to the Sense.* New York: Farrar, Straus, and Giroux.

Darwin, C. (2015). *The Formation of Vegetable Mould Through the Action of Worms, with Observation of Their Habits.* New York: Dossier Press. (Orig. 1882).

Dusek, V. (1966a). Continental Continuum Theory. https://www.researchgate.net/publication/277719895_Continental_Continuum_Theory

Dusek, V. (1966b). The Hamiltonian Revolution in Physics. https://www.researchgate.net/publication/273774695_The_Hamiltonian_Revolution_in_Physics

Dusek, V. (2000). Aristotle's Four Causes and Contemporary 'Newtonian' Dynamics. In D. Sfendoni-Mentzou, J. N. Hattagiani, & D. Johnston (Eds.), *Aristotle and Contemporary Science* (pp. 81–93). New York: Peter Lang Publishers.

Fuller, S. (1997). *Science.* Buckingham: Open University Press.

Fuller, S. (2000). *Thomas Kuhn: A Philosophical History of Our Times.* Chicago: Chicago University Press.

Fuller, S. (2004). *Kuhn vs. Popper: The Struggle for the Soul of Science.* New York: Columbia University Press.

Fuller, S. (2006a). Humanity as the Site of Ideological Conflict in the Twenty-First Century. *Ludus Vitalis, 14*(26), 227–231.

Fuller, S. (2006b). *The New Sociological Imagination.* London: Sage Publications, Ltd.

Fuller, S. (2006c). *The Philosophy of Science and Technology Studies.* New York: Routledge.

Fuller, S. (2007a). *New Frontiers in Science and Technology Studies.* New York: Polity Press.

Fuller, S. (2007b). *Science vs. Religion: Intelligent Design and the Problem of Evolution.* Cambridge: Polity Press.

Fuller, S. (2008). *Dissent over Descent: Intelligent Design's Challenge to Darwinism*. Cambridge: Icon Books.

Fuller, S. (2009). *The Sociology of Intellectual Life*. London: Sage Publications.

Fuller, S. (2010). *Science*. Durham: Acumen Publishing Limited.

Fuller, S. (2011a). *Humanity 2.0: What It Means to Be Human Past, Present and Future*. London: Palgrave Macmillan.

Fuller, S. (2011b). The Posthuman Challenge to Ecological Correctness. https://www.youtube.com/watch?v=r_lzFSnxYZA. https://www.youtube.com/watch?v=UQJ_mWC4dpc

Fuller, S. (2011c). Evidence, What Evidence? *Philosophy of the Social Sciences, 41*(4), 567–573.

Fuller, S. (2014a). *Knowledge: The Philosophical Quest in History*. New York: Routledge.

Fuller, S. (2014b). Social Epistemology: The Future of an Unfulfilled Promise. *Social Epistemology Review and Reply Collective, 3*(7), 29–37. http://wp.me/p1Bfg0-1wG

Fuller, S. (2014c). Keynote Lecture: Russian 'Cosmism' as a Potential Inspiration for Twenty-First Century Interdisciplinary Work. https://www.dur.ac.uk/russianinterdisciplinarity/abstracts/keynote/

Fuller, S. (2014d). Towards a Proactionary Welfare State. *Social Epistemology Review and Reply Collective, 3*(5), 82–84. http://wp.me/p1Bfg0-1sx

Fuller, S. (2015). Retrieved January 28, 2017 from https://twitter.com/professorstevefuller/status/660426459547242497

Fuller, S., & Collier, J. (2004). *Philosophy, Rhetoric, and the End of Knowledge: A New Beginning for Science and Technology Studies* (2nd ed.). Mahwah: Lawrence Erlbaum Associates, Publisher.

Gould, S. J. (1983). A Worm for the Century. In*Hen's Teeth and Horse's Toes* (pp. 120–133). New York: Norton.

Gould, S. J., Raup, D. M., Sepkoski, J., & Schopf, J. M. (1977). The Shape of Evolution: a Comparison of Real and Random Clades. *Paleobiology, 3*(1), 23–40.

Grayling, A. C. (2008, September). Origin of the Specious. *The New Humanist*. https://newhumanist.org.uk/articles/1856/origin-of-the-specious

Hull, D. (1988). *Science as Process: An Evolutionary Account of the Social Development of Science*. Chicago: University of Chicago Press.

Jonas, H. (1967). Judaism, Christianity and the Western Tradition. *Commentary*, (41), 61–70. https://www.commentarymagazine.com/articles/judaism-christianity-the-western-tradition/

Kitcher, P. (2007). *Living with Darwin*. Oxford: Oxford University Press.

Kitzmiller, T. (2005). The Middle District of Pennsylvania: Tammy Kitzmiller, et al.: Case No. 04cv2688 Plaintiffs : Judge Jones v. Dover Area School District, *Tammy Kitzmiller, et al.* v. *Dover Area School District, et al.* (400 F. Supp. 2d 707, Docket no. 4cv2688).

Kuhn, T. S. (1970). *The Structure of Scientific Revolutions* (2nd ed.). Chicago: University of Chicago Press.

Levi-Leblond, J.-M. (1975). Ideology of/in Contemporary Physics. In S. Rose & H. Rose (Eds.), *The Radicalization of Science* (pp. 136–175). London: The Macmillan Press.

Lloyd, E. A. (1994). *The Structure and Confirmation of Evolutionary Theory*. Princeton: Princeton University Press.

Lucretius Titus Carus. (2008). *[94 BCE] De Rerum Natura: On the Nature of Things* (trans. Slavitt, D.). Berkeley: University of California Press.

Lynch, W. (2016). Social Epistemology Transformed: Steve Fuller's Account of Knowledge as a Divine Spark for Human Domination. *Social Epistemology Review and Reply Collective*. https://social-epistemology.com/2016/05/26/social-epistemology-transformed-steve-fullers-account-of-knowledge-as-a-divine-spark-for-human-domination-william-t-lynch/

McHale, D. (1985). *George Boole: His Life and Work*. Dublin: Boole Press, Ltd.

Mesthene, I. (1967). Technology and Wisdom. In I. Mesthene (Ed.), *Technology and Social Change* (pp. 109–115). Indianapolis: Bobbs-Merrill.

Raup, D., & Gould, S. J. (1974). Stochastic Simulation and Evolution: Toward a Nomothetic Paleontology. *Systematic Zoology, 23*(3), 305–322.

Rifkin, J. (1983). *Algeny*. New York: Viking Press.

Schroedinger, E. (1944). *What Is Life? The Physical Aspect of the Living Cell*. Cambridge: Cambridge University Press.

Schweber, S. (1982). Demons, Angels, and Probability: Some Aspects of British Science in the Nineteenth Century. In A. Shimony & H. Feshbach (Eds.), *Physics as Natural Philosophy: Essays in Honor of Laszlo Tisza* (pp. 319–363). Cambridge: The MIT Press.

Segerstrale, U. (2000). *Defenders of the Truth, the Battle for Science in the Sociology Debate and Beyond*. Oxford: Oxford University Press.

Sivin, N. (1995). On the Limits of Empirical Knowledge in the Traditional Chinese Science. In *Science in Ancient China: Researches and Reflections*. Brookfield: Variorum.

Snyder, L. J. (2011). *The Philosophical Breakfast Club: Four Remarkable Friends Who Transformed Science and Changed the World*. New York: Broadway Books.

Toulmin, S. (1963). *Foresight and Understanding*. New York: Harper Torchbooks.

Truesdell, C. (Ed.). (1960). The Rational Mechanics of Flexible or Elastic Bodies 1688–1788. *Leonardi Euleri. Opera Omnia. XI*, Part 2, 288.

Van Fraassen, B. (1991). *Quantum Mechanics: An Empricist View*. Oxford: Oxford University Press.

Weinberg, S. (2007). *Facing Up: Science and Its Cultural Adversaries*. Cambridge, MA: Harvard University Press.

White, L. (1967). The Historical Roots of Our Ecological Crisis. *Science, 155*(3767), 1203–1207.

Wiener, N. (1964). *God and Golem Incorporated*. Cambridge, MA: The MIT Press.

CHAPTER 6

Fuller, Cosmism, and Gnosticism

Abstract This chapter is on an exploration of Fuller's version of Cosmism. This movement, based in part on the Russian Orthodox concept of *theosis* as moving toward a union with God, advocates space travel and the scientific pursuit of immortality. This resembles Fuller's humanity 2.0. There are charges of Gnosticism, which is the Christian heresy holding that the creator of the world was an evil creature and God is beyond this realm and with knowledge (gnosis), one can move beyond this world to a higher form of being, against Fuller's transhumanism. Fuller has been accused of Gnosticism, though the Gnosticism of which he is accused is really eschatology. Fuller himself criticizes biologists who publicly deny biological race differences of being Gnostics. This view is critically analyzed.

Keywords Cosmism • Thesis • Gnosticism

Just as Fuller has drawn and developed many parallels between theological debates and debates concerning science studies (see Chap. 2), so he has related his ideas concerning transhumanism to two movements, Cosmism and Gnosticism. The first of these derives from the Eastern Orthodox branch of Christianity. The second is a heresy denounced by the early church fathers, accusations of which have resurfaced in twentieth and twenty-first century debates.

F.X. Remedios, V. Dusek, *Knowing Humanity in the Social World*,
https://doi.org/10.1057/978-1-137-37490-5_6

Fuller combines his theological positions on humans as the image of God with his transhumanism. Fuller's transhumanism is the view that humanity can be enhanced or redesigned through technology. With converging technologies, which are biotechnology, nanotechnology, and computer technology, humanity can be transformed to an enhanced version of humanity—humanity 2.0. Fuller's transhumanism posits that humans will be transformed through technology to humanity 2.0, into beings that are Godlike.

In his major book on transhumanism, *Humanity 2.0* (2011), Fuller referred to Russia solely in terms of Communism, the Cold War, and the decline of the communist ideal. In his follow-up book, *Preparing for Life in Humanity 2.0* (2013, p. 40, 55, 76), Fuller has emphasized the Eastern Orthodox doctrine of *theosis* to become like and even one with God, based on Jesus' transfiguration. A year later he incorporated the Russian Cosmism movement, in pre-revolutionary Russia, the USSR, and in post-Communist Russia into his reflections on transhumanism (2014a, b; Fuller and Lipinska 2014). Cosmism is a view strongly resembling Fuller's transhumanism in many respects. He has written on and given talks on Cosmism, saying:

> ...perhaps the most signature contribution [of Russian thought to the West] – and the one with the longest half-life – may be 'Cosmism', a heterodox development of Russian Orthodox Christianity, stemming from the philosopher Nikolai Fedorov, who argued on theological grounds that science should aim to make all humans – both living and dead – immortal by acquiring control of the mechanisms of heredity. (2014b)

Elsewhere Fuller has explicitly related Cosmism to transhumanism:

> In retrospect, Cosmism provides the clearest anticipation of today's transhumanist aspirations. It also helps to explain Russia's cultural receptivity to the most ambitious Soviet schemes for world domination, notwithstanding the regime's official atheism. (Fuller 2014a)

Russian Orthodox religion has a concept different from those of Western Catholicism, namely, *theosis*, which involves the person through moral development fusing to a certain extent with God. It also involves the cooperation of human effort and the potentials of God. Greek novelist Nikos Kazantzakis' *The Saviors of God* (1960) is a non-orthodox portrayal of this Eastern Orthodox doctrine.

Fuller writes:

> Arguably the only branch of Christianity that has managed to keep humanity's nascent divinity within a strong clerical context is the Eastern Orthodoxy, which has made the Transfiguration – the moment Jesus realizes his divine nature as "Christ" (*theosis*) – a defining episode in the religion. Yet, even in this case, it is noteworthy that a school of late-nineteenth and early-twentieth century Russian Orthodox thinkers known as "Cosmists" tried to leverage their faith into a hyperbolic scientific worldview that envisaged the physical resurrection of the dead and absolute human control over the Earth and the Heavens via space travel. (Fuller and Lipinska, p. 27)

Although not itself based on Cosmism, but based on Russian Orthodoxy, another tie between the ascent toward the infinity of God and scientific ideas is the genesis of early Russian interest in mathematical infinity. It is notable that an Orthodox priest, Pavel Florensky, published the first accessible accounts of Cantor's theory of infinite sets in theology journals and inspired through religious conversion the founder of the Russian school of descriptive set theory (Graham and Kantor, Pyman).

The Cosmist movement is a secular offshoot of Russian Orthodoxy. Its founder was Nikolai Fedorov, an eccentric Russian who worked as a librarian. He lived the life like that of an Orthodox saint, sleeping on a trunk and giving away most his clothes and other possessions. Despite the obvious similarity of his way of life and ideals to those of Eastern Orthodoxy, Fedorov was not religious in the literal sense.

Fedorov's (1990) bizarre goal was to resurrect all the dead by reconstituting their bodies, reassembling their atoms. This strange or even grotesque goal led to major scientific developments, both in engineering and biology. Fedorov believed that atoms of the departed dispersed in space needed to be collected for reassembly. To do so required space travel. Fuller writes:

> ...perhaps the Cosmist who had the most lasting impact on Soviet thinking was the astrophysicist Konstantin Tsiolkovsky, an inspiration for Sputnik, who presented interstellar travel as breaching the final frontier of humanity's quest for cosmic consciousness. (Fuller and Lipinska 2014, p. 27)

A disoriented and unhappy, even suicidal, youth, Konstantin Tsiolkovsky, was inspired by and became a disciple of Fedorov, who gave Tsiolkovsky the task for space travel of developing rockets, which he succeeded in doing.

> Tsiolkovsky studied engineering and chemistry with this goal. He succeeded in developing rockets at approximately the same time as Robert Goddard did so in the US. Tsiolkovsky wrote of "the infinite potentiality of mankind" and "the perfection of mankind and its individuals." (McDougall 1985, p. 19)

After the Russian Revolution, Lenin was enthusiastic about Tsiolkovsky's plans for developing space rockets and gave him strong support. James Oberg, the leading American authority on Soviet space travel, says that Tsiolkovsky thought of "almost every…concept of spaceflight operations" (1984, p. 197).

Tsiolkovsky's ideas and plans were spread to Germany, giving birth to the German rocketry that eventually issued in the V2 rocket of WWII. Hans Oberth presented these ideas. Although his book was largely ignored, an eccentric crackpot, Max Valier, wrote and inaccurate popularization of Oberth that captured popular enthusiasm. A 13-year-old Wernher von Braun was inspired to his lifetime efforts by reading Oberth (Neufeld, pp. 24–5). The German rocket men, most famously Wernher von Braun after working for the Nazis, were brought to the USA after WWII in Project Paperclip. The German engineers then became leaders in US missile development and the race to the moon. Thus, Fedorov's strange plan issued via Tsiolkovsky in missiles and space rockets for both sides of the Cold War.

Fuller's contribution to the book *Star Ark* (2017), which deals with the launching of a rocket including humans and animals for permanent residence far beyond earth, supports this Cosmist project. In this piece Fuller seems more sympathetic to some ecological concerns than in his other writings. Fuller discusses ecomodernism, a movement that, unlike deep ecology, for instance, does not reject high technology, and hopes to use this to produce a sustainable earth, a "positive Anthropocene," using energy sources that do not deplete either living things or fossil fuels. Fuller also links transhumanism not only to Russian Orthodoxy but also to two Western Christian movements.

"In terms of religious precedents for transhumanism, two of the oldest Christian heresies…stand out…Pelagianism and Arianism" (Fuller et al. 2017, p. 384). Pelagius claimed that humans were able to save themselves through personal effort, rather than through grace, as St. Augustine, opponent of Pelagius and source of Protestant founder Martin Luther, claimed. Fuller characterizes the two heresies succinctly as Pelagians imagined a "heaven on Earth," while Arians imagined "an Earth in heaven." Fuller links Pelagianism to the activism of transhumanism, while he links the intellectual search for a Godlike knowledge to Arianism (Fuller et al. 2017, p. 384). A descendant of Pelgianism is Aubrey de Grey, whose view of indefinite longevity is about biological enhancements to homo sapiens, and of Arianism is Ray Kurzweil, whose view is that homo sapiens will be able to leave their carbon-based bodies and be uploaded to computers capable of colonizing the universe with its expanding consciousness.

Another offshoot owing to Fedorov's inspiration was the earth science and biology of Vladimir I. Vernadsky. The latter approached the earth as an integrated system. His biogeochemistry, uniting the three sciences in its compound name, biology, geology, and chemistry, was a pioneering integrated environmental and earth science. The British-American polymath, the mathematical and chemical ecologist G. Evelyn Hutchinson, at Yale, with Vernadsky's son, translated parts of Vernadsky's book (Vernadsky 1944). Hutchinson developed this science further in a massive survey of biogeochemistry (Hutchinson 1950). Another concept of Vernadsky's was the biosphere, comprising the whole layer of life-supporting environment on earth. This idea was also brought to US science by Hutchinson (1948b, 1970).

Vernadsky's lectures in Paris were heard by and used by Édouard Le Roy, a religiously liberal, unorthodox Catholic priest with an action-oriented interpretation of dogma (LeRoy 1918). He developed a concept analogous to the biosphere, that of the noosphere, the aspect of the earth consisting of human consciousness. Teilhard de Chardin, a creative Catholic priest and paleontologist, who developed a theory of theistic evolution, is culminating in a generalized concept of Christ. Teilhard was shipped off to China by the clerisy, with the hope of keeping his evolutionary and other semi-heretical ideas from spreading. This backfired with Teilhard's discovery of Peking Man (now *Homo erectus*), further supporting human evolution (Aczel). After posthumous publication Teilhard's ideas spread in liberal Catholicism during the 1960s (Teilhard de Chardin).

Russian Cosmism spread in many directions other than that of the engineering of Tsiolkovsky and the science of Vernadsky. It inspired some of the Russian music incorporating color, such as the music accompanied by color projection by Alexander Scriabin, as well as aspects of Russian symbolist poetry. It also was present, albeit in a disguised form, in Soviet projects other than the space program. In the early Communist movement, the "God builders," Anatoly Lunacharsky, Maxim Gorky, and Alexander Bogdanov, attempted to fuse a religious outlook with Communism. Their views were suppressed in the 1920s but were partially revived during the Khrushchev era. The notion of achieving immortality was partially pursued by the preservation of Lenin's corpse and by various attempted early medical projects involving cryonics (Gray). In contrast it is significant that the Soviet biologist and dissident Zhores Medvedev, after being incarcerated in a mental hospital for his dissent, immigrating to England, was very skeptical of plans for extended longevity, perhaps in reaction to Soviet dreams of physical immortality (Medvedev).

The notion of engineering nature and totally controlling the natural environment in Stalinist Russia fits well with the promethean aspirations of Cosmism. Cosmist ideas, advocates, and projects have continued in contemporary Russia.

The religion of space travel as advocated by Carl Sagan, astronomer and searcher for extraterrestrial life, and Neil de Grasse Tyson, head of the New York planetarium and television celebrity, has considerable resemblance to Russian Cosmism. One article by suggests a strong analogy if not a direct tie between Russian Cosmism and the quasi-religion of space travel of Sagan and de Grasse Tyson (Anderson).

Sagan's former spouse, Lynn Margulis, co-inventor for the Gaia hypothesis, supported the publication of Vernadsky's book on the biosphere, but claims that she only discovered the Russian work only after she had developed her own theories. She also sponsored translation and publication of Russian works on symbiosis. Both partners, Sagan and Margulis, were both of Russian ancestry, but apparently Margulis did not read Russian (Ruse).

A filiation of Cosmism that Fuller probably would not be happy about is some of the impact of Vernadsky on ideas of the biosphere and biogeochemistry in ecology. Fuller is highly critical of ecology as major source of the precautionary principle, advocating extreme caution with respect to technological developments. Fuller opposes this with his proactionary principle, the advocacy of willingness to take risks in technological developments.

Vernadsky's work was originally published in English by a man who was perhaps the leading theoretical ecologist, of the twentieth century, G. Evelyn Hutchinson (Vernadsky 1944). Hutchinson was an Anglican with great respect for medieval art and religion, on which he wrote several pieces (Hutchinson 1953, pp. 31–35; 1954). This contrasts with the many

New Age proponents who have used the biosphere and Gaia as neo-pagan concepts in opposition to Biblical religion. Hutchinson was a proponent and disseminator of Vernadsky's notion of the biosphere and related it to global ecological concerns. Margulis was the co-inventor of the Gaia hypothesis, which, in its most extreme extrapolation, has been used to New Age ecological activists to claim that the earth is a living organism, or even a goddess. A much more modest version claims simply that the atmosphere and bacteria form a single feedback system. In between these two interpretations is the notion of the earth as a natural but non-mechanistic organism. Hutchinson's "Circular Causal Systems in Ecology" (1948a, b) gave a sophisticated version of this notion decades before Margulis. Hutchinson was also an enthusiastic supporter of Margulis.

Peter Harrison (2016) is a major source for Fuller, discussing the role of the notion of original sin in as a source of attitudes that led to modern science. However, Fuller supports the Eastern Orthodox conception that Adam and Eve's eating the fruit of the tree of knowledge in the Garden of Eden was not a sin but a result of their naiveté. However, pre-Enlightenment Eastern Orthodoxy had much less interest in and advocacy of scientific research than did Protestantism and Catholicism.

One aspect of medieval Western Christianity and the Franciscan tradition that went against the promethean aspirations of Cosmism is the denial of genuine creativity to humans. Augustine wrote that, "the creature cannot create" (Panofsky, p. 171). This, of course, was replaced by the renaissance idea of self-fashioning and the romantic idea of genius. Fuller claims that human creativity is modeled on God's creation from nothing.

> "...ecomodernism reveals its theological orientation to be closer to intelligent design than to theistic evolution.... *Creatio ex nihilo* – or "Big Bang" – is not simply a break point" between God and the universe, but "...the deity's creative capacity is a model for human emulation...." (Fuller 2015)

This goes strongly against the medieval Western denial of "creation" to humans as mentioned above. It also seems to go against Fuller's discussion of the recalcitrance to matter to God in his work on Intelligent Design. (See Chap. 5.)

Fuller recognizes the resemblance of his own promethean project of transhumanism to Russian Cosmism: space travel and colonization, the overcoming of illness and perhaps death, and the development of a higher form of human being by use of the converging technologies.

The Charge of Gnosticism Against and by Fuller

A number of writers, most notably the political theorist Eric Voegelin and his followers, claim that Gnosticism is a central, negative feature or the modern world view. For these cultural critics, Gnosticism involves an overweening ambition to transcend ordinary humanity through secret knowledge and produce a utopian state.

Some of Fuller's critics characterize him as a Gnostic for his advocacy of transhumanism, while he in turn characterizes biologist critics of race research as Gnostics in a different sense from the former.

Gnosticism is the name given to several heretical Christian sects of the second century. The members did not call themselves Gnostics. The term, meaning secret knowledge, arose only in the sixteenth century but was applied to the ancient sects only in recent centuries.

Historically there were many heretical opponents of what later became the dominant Roman Catholicism of the Early Middle Ages. Gnosticism is a term later used to lump together one group of heretical movements. The movements differed in many respects. Some totally rejected worldly activity. A few are said to have advocated immorality and libertinage (at least according to the heresy hunters). The movements differed on the nature of the world's creation. Some claimed it was by a transcendent deity, while most held it was created by an evil, lesser demiurge.

Most of what was known about Gnosticism before the twentieth century was due to the early Christian heresy hunters whose position later became standard Roman Catholicism. The discovery of the Nag Hammadi manuscripts (Robinson) and various alternative gospels of the Gnostics (Pagels) showed what positions Gnostics really held. Most, but not all, so-called Gnostics emphasized that the world is a place of evil to be escaped through secret knowledge (*gnosis*). For most Gnostics, the creator of this world is not God of the Bible but a malicious creature. Some Gnostics believed that the God of the Bible is in fact that evil lower creator. Escape from this evil world leads one to the true world and the real, transcendent God.

The evil of Gnosticism was a central theme of the writings of Eric Voegelin, a very conservative political philosopher transplanted from Germany to Louisiana. He identified Gnosticism with what the followers call the "immanentization of the eschaton," that is, the making of the transcendent features of the divine as well as of heaven to be manifest in the material world. Fuller admires Voegelin's work on Gnosticism and refers to his "many profound historical studies of Gnosticism" (Fuller 2006, p. 129). Fuller notes that:

> ...Gnostics do not work within the political establishment because they see themselves as accountable to a higher law to which only their party has access.... (Fuller 2002b, 2006, p. 129)

Voegelin (1954) depended for his characterization of Gnosticism primarily upon the small number of ancient documents available at the time he wrote and upon Hans Jonas' *The Gnostic Religion* (2001 [1934]). The discovery of the Nag Hammadi manuscripts changed our understanding of Gnosticism. Voegelin wrote without knowledge of these manuscripts and did not incorporate these later developments of scholarship into his most influential writings.

Voegelin links Gnosticism with Joachim of Fiore, the medieval eschatologist, whose three ages of the world, the age of the Father, the age of the Son, and the age of Holy Ghost, resemble later, modern theories of history, such as Hegel's. However, Joachim had no connection with the Gnostics of late antiquity. Voegelin then traces the filiation from Joachim to modern Enlightenment and to Marxist doctrines of historical progress toward an ideal state. This comparison is certainly defensible to use Joachim but not to the original Gnostics. Voegelin himself, at the end of his life, suggested that perhaps the term Gnostic should not be used for the views he opposed (Webb, p. 49). It seems that Voegelin's "Gnosticism" is really a kind of secular eschatology. It is the belief that the history of the world will culminate in some major event toward which history is drawn, whether, Enlightenment theorists of progress, the early utopians, or Marx's communism.

Nonetheless, without the use of the word "Gnostic," Fuller's advocacy of transhumanism does have resemblance to the views that Voegelin castigates. Fuller claims that the human mind can ascend to overlap with the mind of God and that humans through nanotechnology, biotechnology, and computers can achieve, in some sense, a superhuman status. Fuller does not wish to totally dispense with the image of God characterization of human nature. Interestingly, Fuller does identify transhumanist Ray Kurzweil as a Gnostic (2007, p. 47). Kurzweil speaks of a singularity occurring when super-intelligent computers are developed. Among other things, human immortality will be achieved through cryogenics and high technology. The final state of this development is the uploading of humans into computers (2005). Fuller notes that this extreme strand of transhumanism would resemble transformation to "pure spirit" of Gnosticism, even though Fuller calls this proposal "materialist" (2016). However, if one considers pure information not the same as physical embodiment (e.g. a formal computer code versus the embodiment of that code in various kinds of hardware, of the genetic code as opposed to physical DNA, a sort of Platonic form), then the resemblance to the Gnostic escape from matter into pure spirit is even greater.

In his writings on science and technology studies, Fuller presented another sort of Gnosticism in a negative light (2006, pp. 129–134; 2007, p. 42, 47). Later Fuller seems to accept the characterization as a Gnostic. In response to the article by Shiffman, "Humanity 4.5," in *First Things* (2015), which accuses Fuller of Gnosticism, he tweeted:

> This is the best theological critique of my version of #transhumanism. (2015)

It seems the later Fuller accepts here both Voegelin's account of Gnosticism and to some extent Shiffman's claim that Fuller is a Gnostic. However, Fuller does not generally characterize himself as a Gnostic in another article (2002a, b) commenting on Lessl (2002) and in Fuller's reply to Shiffman (2016).

In one respect Fuller resembles the strain in Gnosticism that wishes to escape from the material world. Fuller has great animus against ecology and conservation, what he calls environmental correctness. As we saw in his treatment of evolution, Fuller dismisses theorists of evolution such as Darwin and Gould as mere "natural historians." Despite his great interest in and extensive reading in human history, the history of life or of the earth has no interest for Fuller. He rejects the historical component of biology and geology. Further, he advocates, as do many techno-utopians, the escape from earth in rockets to leave behind the pollution and devastation left by humans rather than attempting to mitigate it. Fuller's wish to distinguish humans sharply from animals seems to lead him to, or perhaps derives from, lack of appreciation of the non-human environment.

Another application of the term Gnosticism by Fuller is not to himself, but to biologists who oppose research, or alleged research, on biological race differences, and who oppose biological explanations of society and culture. Here the aspect of Gnosticism that Fuller unsurprisingly considers negatively, given his advocacy of the democratization of science, is the Gnostic elitism of special knowers who are superior to ordinary people.

This description applies, says Fuller, to biologists who strongly separate animal evolution from human cultures. Ironically, Fuller himself also wishes to separate humans as the image of God from animals. However, Fuller objects to Stephen Jay Gould and Richard Lewontin in their criticisms sociobiology's E. O. Wilson. Fuller notes Wilson and Gould specialize in invertebrates, not humans, though Darwin was not an expert on mollusks as Fuller writes. Fuller also rightly objects to the three biologists Gould, Lewontin, and Wilson using social science merely as a source of examples, not of theory.

Fuller writes, "E. O. Wilson is a candidate for the mantle of Galileo" (2007, p. 29). Fuller's sole source cited for the debate with Wilson opposing Gould and Lewontin is a book by Ullica Segerstrale (2000) which, although well referenced, is favorably slanted toward her mentor Wilson and snide concerning his opponents, Lewontin and Gould.

For Fuller, the biologists who oppose Wilson's sociobiology are Gnostics, in the sense that they keep a truth for themselves, but propound another for the masses. This doctrine, the "double truth," appears in ancient times in Plato's plan for elitist rule and was defended by the medieval Islamic philosopher Averroes (Fuller, Science 1997; Thomas Kuhn 2000, pp. 92–93; Popper vs. Kuhn, 2004, p. 22). Here Fuller assumes that all professional biologists as professionals accept that there are significant biological race differences. He claims they then propound a different truth, the non-biological nature of race, for the masses (Fuller, Intellectual 2005, pp. 23–24). In fact, a significant minority of biologists, such as Gould and Lewontin, hold that race is not a coherent or a determining quality. Biologist opponents of sociobiology believe biological accounts of culture are mainly wrong.

Defending researchers who believe in significant biological race differences, Fuller claims the opposition of many biologists to research on race difference is sheer "political correctness" (2007, p. 29). They supposedly do this because of the "Gnostic" assumption that "the temptations of forbidden genetic knowledge are so great..." (2006, p. 134). He neglects to consider the sincere belief by that research on biological racial and ethnic differences is not worth pursuing because it is an irrelevant topic for which the only rationale is to pursue claims of racial ranking. For instance, would investment in research on nose shapes of Jews or Africans be considered a worthwhile scientific project? Fuller, in his social epistemology and in his studies of the management of science, emphasizes the role of social utility and economic investment in the choice of research projects worth financing. Does he think there is a valuable social utility in biological race difference research that outweighs that of alternative projects in human biology?

In keeping with his talent to draw connections between theological positions or debates and debates within and about science, Fuller has drawn our attention to the relevance of Russian Cosmism to transhumanism as well as the relevance of attributions of Gnosticism to his own approach as well as that of elite scientists.

References

Fedorov, N. F. (1990). *What Was Man Created For? The Philosophy of the Com Mon Task*. Bath: Honeyglen Publishing.

Fuller, S. (1997). *Science*. Buckingham: Open University Press.

Fuller, S. (2000). *Thomas Kuhn: A Philosophical History of Our Times*. Chicago: Chicago University Press.

Fuller, S. (2002a). *Social Epistemology* (2nd ed.). Bloomington: Indiana University Press.

Fuller, S. (2002b). Demystifying Gnostic Scientism. *Rhetoric & Public Affairs, 5*(1), 133–157.

Fuller, S. (2004). *Kuhn vs. Popper: The Struggle for the Soul of Science*. New York: Columbia University Press.

Fuller, S. (2005). *The Intellectual*. London: Icon Books.

Fuller, S. (2006). *The Philosophy of Science and Technology Studies*. New York: Routledge.

Fuller, S. (2007). *Science vs. Religion? Intelligent Design and the Problem of Evolution*. Cambridge: Polity.

Fuller, S. (2014a). *From Transfiguration to Transhumanism: Taking Seriously Theology as the Science of God*. ABC Religion and Ethics No. 30, September.

Fuller, S. (2014b). *Keynote Lecture: Russian 'Cosmism' as a Potential Inspiration for Twenty-First Century Interdisciplinary Work*. https://www.dur.ac.uk/russianinterdisciplinarity/abstracts/keynote/

Fuller, S. (2015). https://twitter.com/profstevefuller/status/660426459547242497

Fuller, S. (2016). *Letter, First Things*, January.

Fuller, S., & Lipinska, V. (2014). Transhumanism. *Social Epistemology Review and Reply Collective, 3*(11), 25–29.

Fuller, S., Chioti, R., & Ernstsons, K. (2017). Connecting with the Divine and the Sacred and Becoming Cosmically Conscious. In R. Armstrong (Ed.), *Star Ark*. New York: Springer Verlag.

Harrison, P. (2016). Science, Eastern Orthodoxy and Protestantism. *Isis, 107*(3), 587–591.
Hutchinson, G. E. (1948a). Circular Causal Systems in Ecology. *Annals of the New York Academy of Sciences, 50*(4), 221–246.
Hutchinson, G. E. (1948b). On Living in the Biosphere. *Science, 108*, 587.
Hutchinson, G. E. (1950). A Survey of Contemporary Knowledge of Biogeochemistry III. *Bulletin of the American Museum of Natural History, 96*, 1–544.
Hutchinson, G. E. (1953). The Gothic Attitude to Natural History. In *The Itinerant Ivory Tower* (pp. 31–35). New Haven: Yale University Press.
Hutchinson, G. E. (1954). *Scientific Language and Religious Faith*. New York: National Council of the Episcopal Church.
Hutchinson, G. E. (1970). The Biosphere. *Scientific American, 223*(3), 45–53.
Jonas, H. (2001). *The Gnostic Religion* (3rd edn.). Boston, MA: Beacon Press. (Orig. *Gnosis und Satanic Geist*, Vol. 1 1934, Vol. 2 1954).
Kazantzakis, N. (1960). *Saviors of God*. New York: Simon and Schuster.
Kurzweil, R. (2005). *The Singularity Is Near: When Humans Transcend Biology*. New York: Penguin Group.
LeRoy, E. (1918). *What Is Dogma?* Chicago: The Open Court Publishing Company.
Lessl, T. M. (2002). Gnostic Scientism and the Prohibition of Questions. *Rhetoric & Public Affairs, 5*(1), 133–157.
McDougall, W. A. (1985). *The Heavens and Earth*. New York: Basic Books.
Oberg, J. (1984). *The New Race for Space*. Harrisburg: Stackpole Books.
Segerstrale, U. (2000). *Defenders of the Truth, the Battle for Science in the Sociology Debate and Beyond*. Oxford: Oxford University Press.
Shiffman, M. (2015, November). Humanity 4.5. *First Things*, pp. 23–30.
Vernadsky, V. I. (1944). Problems of Biogeochemistry II (trans. Vernadsky, G.). In G. Evelyn Hutchinson (Ed.) *Transactions of the Connecticut Academy of Arts and Sciences, 35*, 483–517.
Voegelin, H. (1954). *The New Science of Politics: An Introduction*. Chicago: University of Chicago Press.

CHAPTER 7

Proactionary and Precautionary Principles and Welfare State 2.0

Abstract This chapter is on proactionary versus precautionary principles. The proactionary principle stresses risk-taking, while the precautionary principle stresses the need to conserve nature. Fuller explores the futures of the human condition, which includes "transhumanism" and "posthumanism." Starting with the emerging challenges posed by so-called human enhancement sciences and technologies, Fuller has explored alternative futures under three rubrics: the ecological, the biomedical, and the cybernetic. These attempts to re-engineer both our bodies and the environment require substantial re-definitions of social justice and economics productivity, all envisioned within new political orders of Welfare State 2.0 as opposed to Welfare State 1.0.

Keywords Proactionary • Precautionary • Posthumanism

In Fuller's exploration of various futures open for the human condition, he considers ones that embrace "transhumanism" and "posthumanism" with respect to epistemic agency. Starting with the emerging challenges posed by so-called human enhancement sciences and technologies, Fuller explores alternative futures under three rubrics: the ecological, the biomedical, and the cybernetic. For Fuller, the ecological is posthuman, which is to resituate the human as one among many life-forms in a com-

F.X. Remedios, V. Dusek, *Knowing Humanity in the Social World*,
https://doi.org/10.1057/978-1-137-37490-5_7

mon environment, the biomedical is genetic engineering and synthetic biology, and the cybernetic is artificial intelligence and it is silicon based, not carbon based. Fuller is betting on either biomedical or cybernetic. These attempts to re-engineer both our bodies and the environment require substantial re-definitions of social justice and economics productivity, all envisioned within a new political order of Welfare State 2.0.

This chapter is on proactionary versus precautionary principles and Welfare State 2.0. Risk assessment is critical to assess the impact of future science and technology on the future of the human condition. The proactionary principle stresses risk-taking as definitive of the human condition, while the precautionary principle stresses the need to conserve nature. In the face of uncertainty and avoidance of risk, precautionary principle states that a policy or action should do no harm. The Wingspread Declaration states that precaution should be taken even if some cause and effect relationships are not established scientifically in terms of health and the environment (1998). The precautionary principle is the basis of European Union scientific legislation.

The proactionary principle, which was formulated by transhumanist philosopher Max More, encourages risk-taking in which the freedom to innovate technologically is critical to humanity (More 2005). In 2004, the proactionary principle was used in reaction to George Bush's bioethics panel to not federally fund stem cell research. Fuller and Veronika Lipinska aver that the proactionary principle, which is an essential part of the human condition, should be favored over the precautionary principle, which is about risk aversion (Fuller and Lipinska 2014). They argue that the proactionary principle provides intellectual basis for transhumanism. They assert that the proactionary principle will unite libertarians and technocrats for a re-imagined welfare state that would reward citizens for risk-taking and protect them if risk-taking leads to negative consequences. They frame the debate on precautionary and proactionary in terms of what it means to be human, transhumanism, and risk-taking. Principles are about axiology/values, and the precautionary and proactionary principles are opposed to each other. Fuller and Lipinska argue that, while the precautionary principle wants to return humans to their biological roots, the proactionary principle wants humans to transcend our biological roots and nature. In terms of an existential response to uncertainty—the precautionary principle is a threat and the proactionary principle is an opportunity.

Though the proactionary versus the precautionary principles are about risk assessment, they are also about epistemic agency, which is on the responsibility of the knower for action. There is a correlation between proactionary and precautionary principles and agent- and object-oriented epistemologies. As cognitive economists, agent-oriented social epistemologists are proactionaries, who take and advocate for more risks. Agent-oriented social epistemologists construct their worlds and promote risk-taking. Object-oriented social epistemologists are precautionaries, who avoid risks. Object-oriented social epistemologists take their worlds at face value, which they perceive to be full of danger and hence avoid risks (see Chap. 3).

One of the keys to Fuller's humanity 2.0 is indifference to material instantiation, including the human body, so long as the person, whom Fuller defines legally as a person, is intact (Fuller 2016). Machines are candidates for humanity 2.0 if they pass the Turing test for self-determination in terms of setting short-, medium-, and long-term goals. According to Fuller, intelligence is artificial and does not depend on biology. An intelligent organism pursues simultaneously a bounded version of all three strategies—seek to win, avoid losses, and stay in the game. Simon's unit of analysis is system of administration, which can be a computer network. This is where AI is relevant (Simon 1981).

Ethics for Humanity 2.0

As questions of ethics for the precautionary versus proactionary principles arise, Fuller and Lipinska view that the proactionary versus precautionary principles are about axiology, and Kant's "ought implies can" principle comes into play here. With nanotechnology and synthetic biology, Fuller argues that in the transformation from humans 1.0 to humans 2.0, the "can" in Kant's principle is not limited by our biological nature. By inserting chips into the brain and with synthetic biology, what is humanly possible has increased. Fuller holds that for superutilitarianism, humans 1.0 with a limited capacity for utility will be replaced with humans 2.0, who have a superior capacity for utility (Fuller 2013). Collin's criticism is that there is a lack of a firm axiological basis for normative guidance of moral action, because the biological substrate of humans 2.0 is radically different from humans 1.0 (2013). For Collin, it depends on the humans we want to create, for instance through editing the human genome, to have moral qualities such as the compassionate Christian or the peace-loving hippy (Collin 2013). Collin argues that Fuller's superutilitarianism requires a proper anthropological base. Though ethics is not reducible to our biological substrate, since our moral judgments are reflexive, ethics depends on our biological substrate. We suggest this is where ethics for Humanity 2.0 is to be considered. Based on Collin's criticism of Fuller, who holds to superutilitarianism, it is not the appropriate normative theory for Humanity 2.0.

Let us consider whether virtue ethics is appropriate. An alternative to Fuller's utilitarianism is virtue ethics, which is not an abstract moral theory such as utilitarianism or deontology, but emphasizes interrelatedness of agents and commitment to values to a community beyond the rules institutionalized in a state. For virtue ethics, morality is not a preference or constraint, it is internalized in practice based on seeking the good life. Nietzsche's views of utilitarianism as a slave morality (every moral agent has to maximize the greatest good for greatest number). Though Nietzsche's view of morality is perspectival, it is compatible with virtue ethics. Virtue ethics would not be appropriate because it is about moderation, while enhancements for transhumanism are to go beyond the human that may be considered immoderate. Deontology is also inappropriate for Humanity 2.0 because deontology is based on duty. Enhancement can hardly be considered a duty in itself though this is what Fuller seems to claim.

Welfare State 2.0

The previous section argued that ethics for humanity 2.0 is inadequate. This section argues that there is an ethics from humanity 2.0, which is the redistribution problem from humanity 2.0 to humanity 1.0. The proactionary welfare state encourages risk-taking and cultivation of human potential. Welfare State 1.0, which protects people from harm, follows the precautionary principle. With a different welfare calculation, Welfare State 2.0 promotes risk-taking instead of protection of individual human lives. Fuller's Welfare State 2.0's three principles are that: 1) when Welfare State 1.0's social safety net is removed, Welfare State 2.0 favors rewards for risky; 2) value preferences change with new technology, and 3) instead of discouraging risky experiments, people should be encouraged with compensation for victims of failed experiments (Fuller 2014).

Fuller challenges the liberal welfare state, which includes protection of individual life. Welfare State 1.0 is influenced by Rawls' liberal notion of citizenship, which is commitment to an individual's freedom, equality of opportunity, and distribution of resources, for a redistributive welfare state. For Fuller's Welfare State 2.0, the proactionary welfare function is calculated differently to embrace risk as flourishing rather than to protect individual lives. Decisions are made individually and collectively. Fuller's Welfare State 2.0 is based on a republican theory of citizenship, which encourages participation in decision making instead of the liberal notion of rights (Remedios 2015).

Fuller's view is that not all capacities in humans 2.0 are enhanced, so there is no master race. As Fuller's view that humanity is immaterial, a computer machine with super-intelligence can be built, which, according to Bostrom, can kill humans (2014). Fuller's view is that there is little chance for a super-intelligence to kill us since what we fear is humanity disappearing to become robots (Remedios 2015).

Objections

There will be the inequality between humans 2.0 and humans 1.0 of humanity 2.0. Why should humans 2.0 redistribute welfare to humans 1.0? Pedersen (2013) follows Collin's (2013) criticism of the lack of a firm axiological basis for normative guidance of moral action because the biological substrate of humans 2.0 is radically different from that of humans 1.0. Fuller advocates for Welfare State 2.0 based on the proactionary principle. Welfare State 2.0 includes hedgenetics in which there can be experimentation involving risks securitized from failed medical experiments on human subjects so that there is insurance for those who suffer from those experiments. Fuller advocates that scientific experiments, even though they could harm human subjects, should not be stopped, because scientists may gather scientific knowledge which may improve mankind. Is this the kind of science that should be advocated? Do long-run arguments work with the possibility that some scientific experiments may destroy humanity? If the destruction of humanity can be a result of scientific experiments, why should humanity accept this? Fuller's response to this objection is he is offering another level of humanity, transhumanism, which has the aim of invulnerability or infinity (Frodeman 2015) that will not require humanity's biological substrate, since humanity can be uploaded to computers. This is where we suggest that the notion of humanity's vulnerability (Coeckelburgh 2013) enters. Humanity has many dependencies making it vulnerable. If we upload our minds to computers, as Kurzweil suggests, there are threats in the virtual world, such as a software program where the virtual world's platform may be the subject of a cyber-attack and in which the Internet is dependent on servers around the world that can be damaged. Even Kurzweil acknowledges that if we are capturing an individual's mental processes, our thinking is directed to physical needs of the human body, and humanity 2.0 will need virtual bodies in nanotechnology-based virtual environments (Kurzweil 2005). There could be new vulnerabilities to virtual bodies in nanotechnology-based virtual environments, such as viruses. Fuller's transhumanists living in the infosphere (Floridi 2002) are relational beings who have vulnerabilities due to relations to other transhumanists (Remedios 2016).

Suffering Smart

As humanity 2.0 pushes against boundaries of ethics, Fuller links theodicy to humanity 2.0. Theodicy is the problem of evil in a world created by an all good God. Fuller's answer to alleviating suffering, which occurs with natural disasters or human deeds, is to suffer smart. He recommends moral entrepreneurship, which is to recycle evil into good through an agent who did evil deeds but has decided to do good. Fuller's examples of moral entrepreneurs are Jeffrey Sachs, George Soros, and Robert McNamara. McNamara, who was the US Secretary of Defense during the Vietnam War, later became the President of the World Bank. He lent money to the Third World to reduce poverty. His lending policies to the Third World had a negative impact because many poor countries were unable to repay their loans and there was corruption in some governments who were recipients of loan money (Remedios 2013).

Problems with the Proactionary Principle

In terms of suffering smart and Welfare State 2.0, this section discusses problems with the proactionary principle. Fuller and Lipinska advocate people voluntarily taking risks for the advancement of technology.

They are critical, in part, rightly so, of institutional ethics review boards. They particularly object to the boards' restriction on experimentation of human subjects. However, as Carl Elliott notes, "If the authors are aware how their plans might sound to vulnerable populations, to disabled people or ethnic minorities, they don't give much evidence of it. Yet it was in response to abuses of these populations that we developed the current research regulation" (2014).

In a libertarian mode, Fuller advocates allowing persons in possession of their faculties to perform dangerous experiments on themselves. This seems justifiable. For instance, in the beginning of the twentieth century, the two biologist Haldanes, John Scott and his son John Burton Sanderson (J.B.S.), seemed to have had a family tradition of doing risky self-experimentation. The former used himself as a subject to study of compressed air illness of divers (the bends) (Boycott et al.). The elder also introduced to famous "canary in the coal mine" as a replacement for the candle and as a warning concerning carbon monoxide (Eschner, Sonne). The latter visited the battlefield in WWI to investigate German poison gases, also investigating gases involved in mining disasters, experimenting on himself by inhaling carbon monoxide. His research paid off in the invention of the gas mask and description of the warning signs of mine gases. The father also experimented on the son. The son experimented by entering a decompression chamber, suffering burst eardrums. His purpose was to aid sailors trapped in disabled submarines. Both Haldanes contributed to the safety of military personnel and of deep sea divers.

A more recent instance of self-experimentation reported in *Wired* is a brain scientist who voluntarily inserted an implant in himself (Engbers). At first things seemed to have gone wrong. Headaches and confusion ensued, but eventually the implant turned out to be harmless and allegedly helpful. Certainly Dr. Kennedy's experiment can be seen by transhumanists an important step in mental self-modification.

Fuller and Lipinska go much further than advocating voluntary self-experimentation, advocating performing dangerous medical experiments on others. They say that damages can be compensated. However, it is doubtful that death or severe mental deterioration could be adequately compensated. They also claim that people should be willing and eager to submit to such experiments for the progress of humankind.

The Nazi medical experiments during post-WWII led to the Nuremberg Code concerned the misuse of medical experiments, followed by the Helsinki Declaration in 1964. The now infamous Tuskegee experiment on uneducated and misinformed African American men misled them into thinking they were being treated for syphilis, but left them untreated, even after treatment was available. The men's death was recorded up into the 1970s until the experiment was exposed (Reverby). The Tuskegee study led to the introduction of restrictions on human experimentation and institutional review boards.

Fuller and Lipiska, surprisingly, do not give a detailed and thorough evaluation and critique of institutional review boards (which in their references appears in scare quotes) given that they seem to be contemptuous of institutional review boards opposing the experiments Fuller and Lipinska advocate. They do claim that such boards, had they been in place in the past few centuries, would have greatly slowed science down (108). Such boards stand in the way of dangerous experiments on human subjects, except in exceptional cases, such as when the patient is terminal, and other medical cures are lacking.

Obviously, medical review boards are often guilty of misjudgment and of secrecy. The boards often serve merely as protective means to avoid future lawsuits by patients or psychology undergraduates rather than for protecting experimental subjects' rights or their physical and mental well-being. However, wholesale condemnation of them or cavalier dismissal of their usefulness is unjustified (Klizmann).

Despite its worthy aims, the notion of informed consent for medical practice and experiments can be problematical (Corrigan, O'Neil). There has been much debate in medical ethics has concerned the characterization of "informed consent." Is the patient or experimental subject truly informed, especially if the medical framing and jargon is not fully understood by the consenting patient? Are the patients or subjects really consenting if they are in a situation of weakness and dependence?

Prisoners who volunteer for dangerous medical experiments are obviously in a highly constrained and dependent position. Rewards of lessened sentences or better prison conditions for the volunteer are in the context of almost total control of the prisoner's options. Many of the "atomic soldiers" (Rosenberg) who walked into the fallout immediately after a nuclear test blast presumably did so willingly, or at least were unresistant, but this was largely because of misinformation supplied by the army as to the relative harmlessness of nuclear radiation. In operation Upshot-Knothole, the officer volunteers participated "democratically," for instance, deciding how far to place themselves from the blast, though doing so unaware of the effects of such a blast (Wellsome, p. 265). Prisoners, who allowed themselves to be irradiated for money, set the levels of radiation for each other, an example of citizen participation in science, though they were unaware of the cancer that would be induced. During the early nuclear arms race, soldiers, prisoners, hospital patients admitted for various illnesses, pregnant women (who became cancer patients and whose children died of cancer generally at age five to eleven) (Wellsome, pp. 219–221), and disabled children were fed plutonium and other radioactive materials. Many adults and children died. The subjects in some cases volunteered willingly. One soldier was misinformed that he was receiving a cure for cancer (that he didn't have). Children fed with radioactive calcium were told they were joining a "science club" (Wellsome, pp. 231–236).

There are prisoners who are guilty of violent crimes who have not only consented to promised brain operations which were claimed to cure their excessive sexual urges (Mark and Ervin). The prisoners hoping for an operation actually threatened doctors who themselves refused to perform such operations. This was even encouraged by the brain scientists who promised the miraculous results from the operations. Showing their authoritarian approach, the same brain scientists in question later advocated quarantine of AIDS victims on Penikese Island, a former leper colony offshore from Boston (Parmet).

Mildly mentally ill patients, desperate for relief or a cure, assented to LSD experiments by the leading psychologist Dr. D. Ewen Cameron, who was surreptitiously working with the CIA's MK-ULTRA program to supply the results, for use in brainwashing (Collins). Cameron then did such experiments as threaten a patient under LSD with a knife, subject patients to a drug-induced coma, and use disturbing, loud, and grating sounds and flashing lights, similar to and foreshadowing the treatment of Soviet dissidents, prisoners in Abu Ghraib in Iraq, and Guantanamo Cuba. Ironically, Cameron had done important work on culturally rehabilitating post-Nazi German society. (Incidentally, the Canadian Psychiatric Association did not rescind his prestigious award despite the revelation and condemnation of his experiments.)

Another controversial experiment was Laud Humphreys' study of anonymous gay sex at a public rest stop (1975). More discussions in professional social science circles have centered on the immorality of the experiments rather than on the lessons learned concerning subordination to authority or the nature of anonymous same-sex behavior (Humphreys, Zimbardo, Milgram), showing how concern over the misuse of experimental subjects has overridden discussion of what was learned from the experiments.

The US government, because of long-standing opposition to marijuana use, legally banned not only the use but also scientific experimentation on the effects of marijuana. This certainly stood in the way of research on medical alleviation of pain in cancer patients. Similar opposition to LSD led to the suppression of experiments that did not support certain claims about its danger. One member of the Edinburgh School of science studies was led from biology research to the sociology of science when he found that his results showing that the then widespread claim LSD damages chromosomes was false, and journals refused to publish the result.

There was valuable formal and informal self-experimentation on LSD. Albert Hoffman, the discoverer of LSD, took famous bicycle ride under the influence of the drug and continued to take it. Sir John H. Gaddum at Edinburgh self-administered LSD and hypothesized that its effects were on serotonin. This positively contributed to a flood of later research and therapy using psychotropic drugs that affect serotonin (Cozzi).

Given the government suppression of research on psychedelic drug, it is understandable that some surreptitiously engaged in formal and informal experiments with such drugs. Although peyote was not banned in the USA until 1964, thereafter, experiments were illegal. Timothy Leary and Harvard ran experiments with psilocybin and later LSD. Andre Weil, a student, exposed their informal experiments and relations with students. Ironically Weil himself was involved with LSD and later became a health food entrepreneur (Lattin). Leary and Alpert were later terminated and segued into mysticism in the case of Ram Dass (aka Alpert) and radicalism in the case of Leary. With the different government and academic opposition to LSD research, such investigation went underground, leading to lack of public and scientific knowledge about psychiatric benefits and psychological dangers of LSD.

Other Harvard psychology experiments, misleadingly presented to their undergraduate subjects, had very negative consequences. Henry Murray, a famous psychologist, who had surreptitiously worked for the Office of Strategic Services and whose experiments were considered to be useful in evaluating people on their resistance to brainwashing conducted the experiments. Students' beliefs were attacked globally, their values denounced and undermined. Many, but not all students, found the experiments highly disturbing. The student most disturbed by them, by the experimenters' own evaluations, was Theodore Kaczynski, later better known as the Unabomber (Chase). One might say Kaczynski was converted by this experience to a strong proponent of a very strong precautionary principle, both respect to psychiatrists and to technology in general.

The Nazi experiments during the holocaust were obviously performed on unwilling subjects. The provocative Fuller defends eugenics and thinks it should not be rejected though stigmatized because of its application by the Nazis. Fuller defends the use of Nazi medical experiments, arguing that the scientific results should not be rejected simply because of their immorality. It has been claimed that testing of persons placed in frigid water to test the limits of endurance can be of use in evaluating the prospects and possible countermeasures of drowned pilots in the arctic waters. However, it has been argued by some doctors that the experimental results are highly flawed. The right measurements were not made. Several results reported are not consistent with other results on hypothermia victims. Several claims about medical effects are erroneous (Berger). Nevertheless, if the experiments really do contribute to scientific knowledge, the ethical and utilitarian issues remain.

Cruel experiments that did have eventual medical payoff were those concerning the testing of artificial blood plasmas on prisoners of war during WWII, for the data of which US intelligence services secretly made exceptions of the experimenters from war crimes prosecution. Several of the former war criminal experimenters started a blood bank and blood plasma corporation in Japan (Yoxen, p. 101) that later accidentally infected many with AIDS. The vast majority of Nazi medical "experiments," such as the amputation of women's breasts, not because of cancer, but to record their reactions, in the concentration camps, were medically worthless. (The resulting gruesome films are on view in the Holocaust Museum in Washington DC.) Another example of useless but murderous Nazi experimentation was Dr. Mengele's disemboweling a pair of twins during a dinner table argument concerning whether the twins' inner organs were symmetrical.

Fuller and Lipinska justify harmful and even fatal experiments, in the cause of human progress and transcendence. One wonders where Fuller and Lipinska themselves would volunteer for such experiments in the service of science and "walk the walk rather than talk the talk," a phrase Fuller has used concerning STS.

A TV advertisement ran in the 1980s saying that Americans are risk-takers, not risk avoiders. Reference was made to the American Western pioneers, their bravery and adventures. This was contrasted with the wimpy contemporary citizens who do not embrace risk. However, the point of the ad was to justify reduction of workplace safety regulations and restrictions on pollution. This shows how the proactionary principle can be used for less than noble ends.

Fuller ties the proactionary principle to what he calls "ecological correctness," (2011) in analogy to political correctness. (Fuller also thinks any rejection of biological human races is based on political correctness, not genetics and anthropology.) Fuller's rejection of Darwinism is explicitly because of Darwin's pessimism and claims that all species go extinct, but apparently tied to Fuller's antipathy to the non-human, non-controlled natural world. One does not have to go whole hog with Fuller's animal rights public intellectual *bete noir* Peter Singer, downgrading the rights of humans to those of other species, to find non-utilitarian value in non-human animals. Fuller characterizes ecological and conservationist attitudes as similar to those of Asian "karmic" religions and opposed to Christianity (where Fuller's Christianity involves a secularized version of eschatology). Presumably Fuller's chapter "Might We become Nazis in Paradise?" (Fuller 2006, pp. 183–195) is in part ironic in praising genuine Nazi ecology initiatives (2006, p. 187, 189), saying "Nazism might well have eventuated in a green paradise," given Fuller's opposition to "ecological correctness" (2011). However, his praise for Nazi preventative health care, German pre-holocaust eugenics of Alfred Ploetz (2006, pp. 187–189; Fuller and Lipinska, p. 95), and some Nazi medical experiments (pp. 108–109) is sincere.

Perhaps linked to Fuller's visiting professorship in Denmark, he defends the eco-skeptic Lomborg, who is a prolific, contrarian public intellectual like Fuller. Lomborg extensively criticizes of ecologists on economic grounds (Lomborg 2001, 2010). Fuller emphasizes a justifiable defense of Lomborg against the inquisition by his university on grounds of falsification or misreporting of data, Fuller saying that it was due to outrage at Lomborg's anti-ecology. However, very many Lomborg's footnotes have been claimed to be either inaccurate, untraceable, or misleading (Friel). Fuller dismisses critics of Lomborg as being mere biologists, not familiar with economics. Fuller's disdain for the scientific ecologist and evolutionist critics, such as Thomas Lovejoy, is strengthened by his advocacy of Intelligent Design. On the other hand, Fuller elsewhere praises evolutionist E. O Wilson as a modern Galileo because of the latter's sociobiology. William Nordhaus, a leading economist who has written extensively on the economic effects of climate change, who is not at all, by the way, an ecological radical, who uses future discounting to downgrade negative effects of climate change, has produced results very different from those of Lomborg's. The latter's claims are in part based on a researcher, Richard Tol, who not only misinterprets or misrepresents Nordhaus but makes simplistic economic calculations to make his case. Also, Tol uses cost/benefit analysis to argue that the few hundred dollars per capita per year saved by Europeans on air conditioning outweigh the small annual incomes lost by the Africans killed by global warming, thereby producing a beneficent result (Edis).

Fuller's solution to the technological depletion of resources and extermination of non-human species on earth is shared by many technology enthusiasts—escape from earth. The StarArk (Fuller et al. 2017) project, with which he is associated, involves building an ark-like spaceship that will transport humans from the polluted and devastated earth. Besides the issue of technological feasibility of colonization of planets beyond Mars and on distant solar systems is the matter of an overloaded StarArk. Just as Noah's Ark would have trouble carrying one million plus species, so Fuller's Ark would be unable to accommodate the whole human population of billions, whose growth is advocated by Fuller, as by Marxist and Roman Catholic doctrine, limiting abortion, euthanasia, birth control, and other population limiting procedures. Perhaps the StarArk would only be able to accommodate a Strangelovian elite of proactionary *Uebermenschen* adventurers, leaving the ecologists, precautionaries, and most other benighted humans to perish.

Fuller's and Lipinska's defense of the proactionary principle challenges many contemporary truisms concerning the ecology movement, medical review boards, and technological development. For both supporters and opponents, Fuller and Lipinska's advocacy raises numerous issues and opens many avenues of inquiry.

References

Bostrom, N. (2014). *Superintelligence*. Oxford: Oxford University Press.

Coeckelburgh, M. (2013). *Human Beings @ Risk*. Dordrecht: Springer.

Collin, F. (2013). Two Kinds of Social Epistemology. *Social Epistemology Review and Reply Collective, 2*(8), 79–104. http://wp.me/p1Bfg0-Ul

Elliott, C. (2014, September 3). More or Less than Human? *New Scientist*.

Floridi, L. (2002). On the Intrinsic Value of Information Objects and the Infosphere. *Ethics and Information Technology, 4*(4), 287–304.

Frodeman, R. (2015). Anti-Fuller: Transhumanism and the Proactionary Imperative. *Social Epistemology Review and Reply Collective, 4*(4), 38–43. http://wp.me/p1Bfg0-1Zr

Fuller, S. (2006). *The New Sociological Imagination*. London: Sage Publications, Ltd.

Fuller, S. (2011). The Posthuman Challenge to Ecological Correctness. https://www.youtube.com/watch?v=r_lzFSnxYZA. https://www.youtube.com/watch?v=UQJ_mWC4dpc

Fuller, S. (2013). *Preparing for Life in Humanity 2.0.* New York/Basingstoke: Palgrave Macmillan.
Fuller, S. (2014). Towards a Proactionary Welfare State. *Social Epistemology Review and Reply Collective, 3*(5), 82–84. http://wp.me/p1Bfg0-1sx
Fuller, S. (2016). *The Academic Caesar: University Leadership Is Hard.* London: Sage Publications.
Fuller, S., & Lipinska, V. (2014). *The Proactionary Imperative: A Foundation for Transhumanism.* New York/Basingstoke: Palgrave Macmillan.
Fuller, S., Chioti, R., & Ernstsons, K. (2017). Connecting with the Divine and the Sacred and Becoming Cosmically Conscious. In R. Armstrong (Ed.), *Star Ark.* New York: Springer Verlag.
Kurzweil, R. (2005). *The Singularity Is Near: When Humans Transcend Biology.* New York: Penguin Group.
Lomborg, B. (2001). *The Skeptical Environmentalist.* Cambridge: Cambridge University Press.
Lomborg, B. (2010). *Cool It: The Skeptical Environmentalist's Guide to Global Warming.* New York: Vintage.
More, M. (2005). *The Proactionary Principle.* http://www.maxmore.com/proactionary.html
Pedersen, D. (2013). Who Should Govern the Welfare State 2.0? A Comment on Fuller. *Social Epistemology Review and Reply Collective, 2*(12), 51–59. http://wp.me/p1Bfg0-19F
Remedios, F. (2013). Review of Humanity 2.0: What It Means to be Human Past, Present and Future. *LSE Review of Books.* http://blogs.lse.ac.uk/lsereviewofbooks/ 2013/05/07/book-review-humanity-2-0/
Remedios, F. (2015). Knowing Humanity in the Social World: A Social Epistemology Collective Vision? In J. Collier (Ed.), *The Future of Social Epistemology: A Collective Vision* (pp. 21–28). London: Rowman and Littlefield International.
Remedios, F. (2016). Steve Fuller: Knowledge, the Philosophical Quest in History. *Metascience, 25*(1), 3–23.
Simon, H. (1981). *The Sciences of the Artificial.* Cambridge, MA: MIT Press.
Wingspread Statement on the Precautionary Principle. (1998). http://www.who.int/ifcs/documents/forums/forum5/wingspread.doc

CHAPTER 8

Conclusion

Abstract This chapter is the Conclusion, which is an overall assessment and significance with how Fuller's work hangs together from 2000.

As a controversial intellectual provocateur, Fuller's writings since 2000 cover many topics, Fuller is criticized as not being systematic in his work. Fuller defends humanity from the impact of the technosciences in his writings from 2000 with his question on what kind of being should the knower be. Fuller's answer to his question before 2000 is that knowledge should be organized normatively through the knowledge policy analyst. Fuller's answer to his question after 2000 is the technosciences transforms the knower from humanity 1.0 to humanity 2.0. In this book, we uncover that the thread to Fuller's work since 2000 is that the epistemic agent makes knowledge through the construction of reality. Fuller's question: what kind of being should the knower be is a reflexive question on what kind of knowers do we want based on the transformation of knowers by the technosciences. Fuller's further question is: do we want Darwinian type of knowledge, which is adaptationist knowledge, or transhumanist knowledge? We read Fuller's version of the ID debate in terms of his defense of humanity against evolutionary theorists. Transhumanism reinvents the prerogatives of the human to be not limited by its biology, and this makes humans to be Godlike. For Fuller, to do science is to participate in the mind of God because humanity is created in the image and

F.X. Remedios, V. Dusek, *Knowing Humanity in the Social World*,
https://doi.org/10.1057/978-1-137-37490-5_8

likeness of God. Scientists, who are knowers, actively participate in the technosciences to transform humanity. Fuller's question on what kind of being should the knowers be differentiates his work from other transhumanist thinkers such as Nick Bostrom, James Hughes, Julian Savulescu, and David Wood, who do not ask this question.

An original way to look at universities as knowledge producers is Fuller's work on universities as corporate agents who produce knowledge for the public good. This also fits into his question: what kind of being should knowers be? Specifically, his question on universities is: what kind of knowers should universities be with the impact of the technosciences as universities are the premier site of knowledge production? Should universities be only involved with research and knowledge to be kept with the researchers? Or should the research knowledge be distributed through teaching to lower the power imbalance of researchers and students to achieve epistemic justice, which is the responsibility to integrate teaching and research. Fuller's answer it is the university's the epistemic agent who integrates research with teaching also transforms humanity. The integration of research and teaching is achieved through the university president, the Academic Caesar. Research and teaching should be integrated because the university is a premier site for knowledge production through research and it has the responsibility to distribute knowledge through teaching (Fuller 2016a). The corporate agency function of the university provides its raison d'être because its mission is to produce knowledge which will benefit the state. For university 2.0, it is the student who has to integrate the capacities of herself for *bildung*, which is the idea of a self-directed and self-formed person.

Though Fuller has written on post-truth as embracing it (2016b), Fuller's view is that science's aim is truth as a systematic representation of reality in terms of a grand unified theory of everything. How is Fuller's view of post-truth reconciled with his view that science's aim is a grand unified theory of everything? Fuller's view of post-truth is a second-order view on setting the conditions of truth. Social epistemology is also about modal power, on what is necessary and possible.

Kuhn's *Structure of Science of Revolutions* (1970) view is that truth is relativized to a paradigm and there is no correspondence of a scientific theory to reality. Fuller's view is that Kuhn's account is post-truth because truth is not a source of legitimation of scientific theories but is a mask of legitimacy for those in pursuit of power. Pareto's explanation of masks is on the maintenance of social order is in terms of Machiavelli's foxes and lions. Foxes gain power through demystification by showing what has happened behind the scenes. Lions are preservers of power through lineage. STSers are like foxes as demystifiers on showing what is happening behind the scenes in science.

On theodicy, Fuller asserts that moral values will become epistemic values in the long run so that the problem of evil diminishes. Fuller's argument is that if one adopts the perspective of the mind of God, the problem of evil is diminished because of the super-utilitarian argument of the greatest good for the greatest number from scientific achievements for in the long run outweighs bad outcomes of science such as experiments on victims of Nazi atrocities. Fuller advocates that scientific experiments with harm to human subjects should not be stopped because scientists may gather scientific knowledge which may improve mankind. Is this the kind of science that should be advocated? Do long run arguments work with possibility that some scientific experiments may destroy humanity? If destruction of humanity can be a result of scientific experiments, why should humanity accept such a result? Fuller's response to this objection is that he is offering another level of humanity, transhumanism, which has the aim of invulnerability or infinity (Frodeman 2015) that will not require humanity's biological substrate since humanity can be uploaded to computers. This is where we suggest that the notion of humanity's vulnerability (Coeckelburgh 2013) enters. Humanity has many dependencies to which it is vulnerable.

As Fuller claims that in doing scientists entering the mind of God, the issue of bounded rationality enters the picture. Bounded rationality is about the constraints to the knower. If the aim of transhumanism is to be infinite, how will transhumanism address bounded rationality? Unbounded rationality is the rationality of omniscient superintelligence, which is God, who is not constrained by limited time, knowledge, or computational capacities. Transhumanism aims to be Godlike. Bounded rationality is based on Herbert Simon's (1991) notions of the limitations of the human mind in that in real-world situations, optimal strategies are unknown or unknowable, and the structure of the environments in which the mind operates can explain why heuristics perform well: if the structure of the environment is adapted to that of the environment (Remedios 2016).

Fuller's response to Remedios's criticism is that Simon is optimistic that a computer can be programmed on second-order discovery process "learning how to learn" to overcome human biases to become a superintelligence and hence be Godlike in its decision making (2016). In today's world of artificial intelligence, computers are programmed to learn how to learn through deep learning programs. But deep learning programs are based on data which have inherent biases. For example, with predictive analytics on crime in a city, the crime data used for projections has biases in the data based on the actions of the police who target groups that have more tendencies toward crime. In the same way, if computers with deep learning programs are to make predictions in science, those predictions are based on scientific data, which have inherent biases. Fuller may respond that the computer can be programmed to look for and correct biases in the data. That is very hard to achieve because altering the scientific data changes the prediction and the validity of the prediction cannot be proofed.

On Intelligent Design, Fuller thinks he must reject evolution to protect the dignity and special place of human agency. His criticisms on purely natural selectionist evolution with emphasis on pure randomness of variation justify this stance. However, biological evolution need not be purely a matter of total randomness and selection. Directed mutation, physical self-organization tendencies, teleological aspects of adaptation, and other factors are other proposed causes of evolution.

Fuller also rejects theistic evolution as a metaphysical idle wheel. This is the case if the pure randomness plus selection account is accepted and God's role makes no difference. However, it is possible to combine a theistic account of evolution with a process that is not solely hyperselectionist. Against standard ID, Fuller wishes to discern the purposes of the creator in the structure of organisms, but this does not exclude doing this, if it is possible in general, along with some forms of evolution.

On transhumanism, Fuller is rightly critical and skeptical concerning many of the pronouncements of scientific spokespeople as purveyors of hype. However, Fuller is totally uncritical about the similarly of utopian technologists' and corporate leaders' positions on artificial intelligence, synthetic biology, and space travel. He assumes computers can replace human investigators and allow the uploading of human thought and personality. However, he never discusses and replies to the technical and philosophical literature that claims there are limits to what is claimed can be achieved toward strong artificial intelligence, or with genetic engineering, and space colonization.

References

Coeckelburgh, M. (2013). *Human Beings @ Risk*. Dordrecht: Springer.

Frodeman, R. (2015). Anti-Fuller: Transhumanism and the Proactionary Imperative. *Social Epistemology Review and Reply Collective, 4*(4), 38–43. http://wp.me/p1Bfg0-1Zr

Fuller, S. (2016a). *The Academic Caesar: University Leadership Is Hard*. London: Sage Publications.

Fuller, S. (2016b). *Science Has Always Been a Bit 'Post-Truth'*. https://www.theguardian.com/science/politicalscience/2016/dec/15/science-has-always-been-a-bit-post-truth

Kuhn, T. S. (1970). *The Structure of Scientific Revolutions* (2nd ed.). Chicago: University of Chicago Press.

Remedios, F. (2016). Steve Fuller: Knowledge, the Philosophical Quest in History. *Metascience, 25*(1), 3–23.

Simon, H. (1991). Bounded Rationality and Organizational Learning. *Organization Science, 2*(1), 125–134.

CHAPTER 9

Postscript

Abstract This chapter is a Postscript, which includes an interview with Fuller that will involve his reflections on the path his social epistemology has taken to date.

Interview with Steve Fuller—August 21, 2014, Buenos Aires, Argentina

Your new book, *Knowledge* (2014b), makes the distinction of agent-oriented and object-oriented social epistemology. In light of this distinction, I am considering situating my forthcoming book with Val Dusek, *Knowing Humanity* (2018), which is based more on agent-oriented social epistemology and the changing boundaries of the knower:

1. Social epistemology's original question is about how knowledge should be organized. In *Social Epistemology* (1988), Fuller seems to come down more on the side of object-oriented social epistemology in terms of organizing disciplines or at least you were ambivalent between choosing either agent-oriented or object-oriented social epistemologies. Why did you change to agent-oriented social epistemology?

Fuller—I do not know if I have changed, maybe it is a change in emphasis. I think you picked on that in original *Social Epistemology* (1988); there was a lot of emphasis on defining the object of inquiry and making it very clear it was not about beliefs in people's heads. This is against classical

F.X. Remedios, V. Dusek, *Knowing Humanity in the Social World*,
https://doi.org/10.1057/978-1-137-37490-5_9

epistemology model of knowledge. In fact I was quite explicit that knowledge is externalized primarily in texts and as publicly available. I take knowledge very much as stuff, things in the world. My view has not changed on that point. However, there may be a change in emphasis that is still consistent with the original vision. First of all, if you recall, *Social Epistemology*, the book was very much preoccupied with the policy maker who organizes knowledge. The book has a strong second-order sensibility about it. That is why the book was marked social epistemology as a knowledge enterprise. In this sense, there is a conception of knowledge policy maker as an the agent in the book. In the book, there is discussion of this person as a human in straightforward terms. There was no discussion on boundaries between the agent and object of knowledge being blurred in the same way. The other thing is constructivism as an underlying epistemological principle, which is consistent in my work; constructivism implies an open-endedness as you are talking about knowledge as a construction, which involves an agent doing something in the world. If I were a classical realist who believes that knowledge is about correspondence with the facts of the matter, I would not require a strong sense of the agent because as long as I have the objects right, the nature of the agent is not important because the objects are a mirror of reality. But I never had that view, my view is the constructivist view, which implies that the agent is making knowledge. So there is always agency. If you look at my interest in rhetoric, it is about making it up on the spot. The contextual nature of knowledge is about this. There was implicitly there was a strong sense of agency. Though there was not theorizing about the boundaries of agency.

FR—My reading is that earlier Fuller was not a strong constructivist in earlier work but a realist in the social sciences. How does strong constructivism match with realism as a metascience?

Fuller—It is a realism about agency because the agent is responsible for his or her actions. To have a normative view of knowledge and to play with the idea of what it is to be a knower and not to have a classical view of homo sapiens which may be a transformed creature, then you still have to keep a clear sense of agency and what is the locus is today. As it were the being that could be held responsible for its actions including its epistemic actions. That is what I would identify realism with is the ability to find the locus of responsibility so that you do not end up getting this highly diffuse view in what you get with actor-network theory, which is responsible for everything and nothing, in its view of agency. So I am not so antirealist that I would allow that to happen. If you cannot ascribe responsibility to agents, it is very difficult to talk language of causation, to talk the language of plans, change, all these kinds of things are so important. If you want to talk about you talk about progress in doing things in the world, you need to have a sense that you doing things what it is they are doing how you can be held responsible.

FR—There was a paper (2003) I wrote on Joseph Rouse, and Rouse's response (2003) was while he views that normativity is all the way down and his question to me and to you was where is causation and your answer now it is with the agent.

Fuller—That is right, I mean it in the back of my mind in this goes back to my PhD days, I actually see the notion of normativity and responsibility, and causation actually quite bound up with each other because the influence of legal thinking had in my work. In the law, you cannot do anything normative unless you can ascribe responsibility for agents who are held accountable for their actions, and these agents may be individuals, they may be corporations, they may be nation-states. In other words, the ontology of the agent is quite a fluid situation, but nevertheless there is a clear notion agent who can be held responsible. In terms of where I could get this idea from ultimately is that I was very much influenced by causation in the law by Hart and Honore and that was from my training in Cambridge and I do talk about that in my PhD. I do not believe in this idea that normative is just diffuse all over the place. I believe you need you actually do need a conception of an agent.

2. Your work on interdisciplinarity and universities as the premier site of knowledge production seems to be on the side of object-oriented social epistemology since it is about how disciplines should be organized. How does agent-oriented social epistemology fit in to them?

Fuller—The university as the years have gone by is where the world has changed. Since 1988, the university itself can no longer be taken for granted as having an ontological integrity so it has to be defended as something in an explicit way, and in this respect I have shifted my focus on the university from a site where the disciplines are organized, which I think is fair to say as it was in *Social Epistemology*. In a sense I was interested in universities in which disciplines were organized, and I was interested in how disciplines are formed and reformed. And that was even true it may be *Philosophy Rhetoric End of Knowledge* (2004), but now the situation with the university is that it has been problematized so much as an entity that one needs to defend the university. I defend the university certainly with works like the *Knowledge Management Foundations* (2002) and *Sociology of Intellectual Life* (2009). In those works, the University is an agent who constructs knowledge as a public good.

FR—So the university itself is an agent?

Fuller—Exactly the university is a corporate agent. This is why I have always been very attracted to the idea corporations as agents. I am very sympathetic to the idea of corporate agency because this is what you need to secure the university's integrity as an institution whose agency entails its ability to produce knowledge as a public good because that is its unique function and it needs agency in order to do that.

FR—The university now is very diffuse and there are so many faculties; it is hard to see where the agency fits in. With so many faculties, how do you get to agency?

Fuller—When people ask me what social epistemologists should be doing, my answer is that they should be in university administration, they should run the corporation. You know at the end of the day, you want to be running the corporation, you will be the manager of the corporation. You want an overview and be the person responsible for keeping the corporation operating with a clear horizon and with a clear focus so it just does not blur everything else in the world. In other words on the university becoming diffuse as an organization, I lay that at the doorstep a poor administration or administrators who do not have a clear idea in their own mind about what the differences between university and some other corporate agent.

FR—They are business people!

Fuller—Exactly there has been a shift but that are what are the consequences that we live in a world where knowledge is very dynamic. Knowledge produces quite a lot of change in the last 25 years.

FR—What about on the interdisciplinary side? How does the agency fit in there?

Fuller—In the keynote I gave at the meeting (2014c) in Blacksburg, I talked about the techno-Goethes, and the idea is there are two ways of thinking about interdisciplinarity. I outline this point in my *Synthese* piece 2012 on deviant interdisciplinarity. You can think of interdisciplinarity as a feature at the domain of knowledge which is organized in a certain way, and interdisciplinarity is about mapping it out and figuring what are the jurisdictions and what belongs where, who speaks for what, it is a cartographic, something the Neo-Kantians were good at. The other kind of interdisciplinarity that I endorsed is deviant kind is where the individual integrates interdisciplinarity in themselves and they constitute their agency in that process so that each agent, each individual who studies the vast array of knowledge integrates into their own personal synthesis which then has a character of their own. This is Humboldt's image of liberal arts education. This is what you should be doing when you get a liberal arts education, and this is kind of model which I call the techno-Goethe now because we need a technological component, and this cannot be seen as the humanities only. This is the interdisciplinarity I am interested in where it becomes constitutive of agency in as the individuals have to organize the disciplines for themselves.

3. In my 2003 book *Legitimizing Scientific Knowledge*, I pointed out that your view of the social epistemologist is a thin notion of agent and you needed a collective of agents or an institution of agents. At that time, you seem to agree. Now with transhumanism, you seem to have a super-agent with Godlike capabilities through extension of human capabilities with science and technology. Is Fuller the super-agent?

Fuller—I do not disagree with the idea of a thin sense of an agent in the sense that it is a part is constructivism point. I mean in that you do not want a notion of agency that is built with a lot of essential features so it cannot be deep in that sense right and this is why I do not care much for virtue theory and stuff like that because I think it comes in too loaded with what is intrinsically right and wrong. It is a way to think on how to behave and it is not sufficiently open to the world. And so in that respect, I still have a thin conception of agent, but a thin conception agent is not necessarily to have less conception of agent. How much ontological depth do you actually need to build into an appropriate conception agency? If you look at the law, for example, with regard to issues personhood, legal identity, if you look at what the law specifies the boundaries, but let us say corporate personhood, it does not go into what a substantive details right, it is mostly detail about the way the entity relates to other entities and about maybe something that the internal structure, the organization terms, its internal accountability, but it isn't there's a lot of substantive assumptions about the nature of corporate entity. So the corporate entity could be a business firm, it could be a church, it could be a university so in that sense agnostic and in that respect thin agency.

FR—So what about Steve Fuller as super-agent?

Fuller—I do think this goes back to the question of what constitutes the knowledge policy maker person. I do think at the end of the day as it were a training in social epistemology would enable you to actually engage in this kind of work so you could be someone who could run an university or you could be advising about knowledge transfer in all sorts context things of that kind so you would with the knowledge be empowered with a certain kind to the agency.

4. Is the difference between agent-oriented and object-oriented social epistemology a difference in degree or in kind? You make it out to be in kind with the masters and doctors distinction. Is it really in kind because how does the agent organize the disciplines without organizing the objects of the disciplines or does the agent organize him or herself, which is the difference in degree?

Fuller—I think it's a more the agent organizing himself or herself, and this is why I am quite sympathetic to German idealistic thinking. German idealism is about the idea that when you have a strong subject object distinction, it is a problem in that it shows that you have something that has not been incorporated into yourself in some way right so that its other to you. Part of the knowing process is the incorporation of knowledge into your being so it becomes part of your world, that is, the general tenor with German idealism was on about and so you know in principle you could take on everything within yourself. This what Goethe think that he was doing and as well Emerson and Hegel. They were very much in the tradition. They become world encompassing in their own being and that is what I am sympathetic to. But I do think the first stage or early stages of this process do involve a kind of inventory of what are the objects of knowledge, what are the things external to me with the different things in the world as they present themselves to me, a kind of phenomenology of academic life as it were. You got natural sciences and social sciences, the neo-Kantian view is you classify all these different things out there. Now the question is do you just leave it like that, this is what science studies does, it gives you this kind inventory, it tells you something about the relationships, but it does not really do anything more with it, it does not do anything second order to organize it, or you know say it is all really the same thing or something like that, but this is what I am into. I am into that second-order project which involves then incorporating those external objects of inquiry into the enquirer.

FR—And then doing something about those objects organized as those objects?

Fuller—It is very difficult, here is a very concrete example, animals, they are running around very much as other to us. We can have a science called biology and what it tells you is the morphology, genetic makeup, the environments in which all these animals live. But then the question becomes what do you do after that is a just a mapping exercise or is there some sense something you want to incorporate that has explanatory value world view so it is important to you that there are these animals out there and this is where the issue of the significance a genetic overlap with ourselves becomes important to the levels at which there is common ground and then is there is some kind of a larger normative understanding that encompasses both us and them. This is why I take an interest in something like animal rights because animal rights basically gives me something that I do not specifically agree with but it does give you that normative closure in some sense the animals are built into our ethical system. If you are Peter Singer with a utilitarian system then there is a welfare function that the covers over animals as well as humans. I do think that this is what you do next, this is what the social epistemologist does, it takes the up the objects of knowledge builds into the agent and then provide the kind of normative cover right that rationalizes the whole process.

5. In Table 9.1 of *Knowledge* (2014a), you hang a lot on the distinction between agent- and object-oriented social epistemology such as transhumanism falls on side of the agent. If the distinction fails, what happens to transhumanism since it is linked to agent-oriented social epistemology? What happens for you if it fails in any way or you do not see it failing anyway?

Fuller—I think first thing to say if you are Bruno Latour or any of these post-modern guys, they really do not want to really recognize this distinction. But I do think in a sense the distinction is a normatively salient one from the standpoint of what you think the ends of knowledge are. Let us put it that way because if you are an object-oriented social epistemologist then the fact that there are experts in the world, the world is divided up

Table 9.1 The existential horizons of social epistemology

Version of social epistemology	*Agent-oriented social epistemology*	*Object-oriented social epistemology*
Nature of knowledge	Knowledge is unified according to our subjective interests	Knowledge is divided according to the nature of its objects
Underlying metaphysics	Epistemology "artificially" constructs ontology	Epistemology "naturally" mirrors ontology
Status of humans	Humans are aspiring deities	Humans are receptive animals
Structure of mind	Will extends the intellect: "anticipation of experience"	Will is subordinated to the intellect: "reflection of reality"
Christian precedent	The dissenting tradition: Franciscans, Protestants	The establishment tradition: Dominicans, Catholics
Locus of academic authority	Masters of Liberal Arts	Doctors of Professions
Temporal authority	The future (unless checked otherwise)	The past (unless checked otherwise)

into these domains that carries a certain normative implication which has to do with the importance of authority and importance of deference and trust. A lot of these kinds of issues in recent times analytic social epistemologists have been emphasizing as normative virtues of knowledge that in the past were not stressed so strongly but are now very important. I think that really does require basically adopting the object-oriented approach because what you are saying here with the expertise is that in some sense the objects dictate what needs to be known. The people who as it were know what needs to be known compared to the object, which is driving the whole enterprise. Whereas if you are an anti-expert you come in as it were with your own aims, your own interests and then you evaluate knowledge by the extent to which it promotes those. From the agent-oriented epistemologist standpoint, the existing expertise that exists in society, the division labor of knowledge could be seen as an obstacle, as a barrier, as a way of obstructing the possibility of getting things done because you have to do all this training in that the you have to master all these domain languages. You have to do all this kind of work.

I do think there are very strong normative implications built into each position it so that is why we keep them distinct. The normativity of what they imply are different. We live in a world that is increasingly expertized and the question is it a good or bad thing? I can think it is bad thing and in any case I certainly do not want it naturalized. This is what I think the problem is with the guys on the object-oriented side. You think you need to have both the social epistemologies on the table so the people see with the differences and they are not just all the same thing.

FR—Is the social epistemologist a type of expert?

Fuller—Not in the sense that we are talking about. It becomes really important to have an open-ended conception of what social epistemology is about because social epistemology is not just another discipline. It is a second-order field. In a way it is a transcendental field in the sense that it is always second order so if other disciplines get invented, social epistemology is above that too. So it is more like a transcendental position or a meta-position.

6. Where is the present in Table 9.1? Agent-oriented social epistemology is about the future and object-oriented social epistemology is about the past. What about the present since we are in the present?

Fuller—I actually think that the present is a temporal mirage. I do think what counts as the present, actually I say this in the last chapter of *Knowledge* (2014a) the one on the counterfactual historiography. I think that what counts as present is very much but dictated by context. For example, if we say something like we live in the modern era that is the present. We are establishing identity for the purposes of this discussion that extends to the eighteenth century insofar we are saying that certain kinds of ideas that were important then are important now. So in the way the identity of present is established by making links both forward and backward. I do not believe in a pinpoint, there is no center is that now and this is where I wish I really do think discourse analysis can be quite interesting because I think the present is always shifting for depending on the context we are talking about. So in some context, for example, in respect of the modern age, I think we assume Kant is our contemporary. But obviously there are a lot of other contexts Kant is not our contemporary as we are talking about a figure from the past, and we are somewhere else.

FR—But also to present includes now so what is happening actually right now?

Fuller—But the now you know that was what now with the era of neoliberalism so everything up to 1989. This is now or is the now the computer age or which now we are talking about. We are not talking about the species present. I mean idea of the species present that there is some magic moment, some ecstatic moment where we are right now for only and only now.

FR—Remember be here now!

Fuller—The uniquely instantaneous present that happens only at the time it happens.

FR—Is there be some hybrid version of agent- and object-oriented social epistemologies or is there not a hybrid version?

Fuller—I make these distinctions at the end of the Introduction in *Knowledge* (2014a). I exaggerate distinctions as a dialectical move in order to make the normative implications very clear. In the course in real life, we do shift between these things, they are what the discourse analysts call réper-toires. Sometimes you talk one way, sometimes you talk the other way right to shift between them. I think the normative implications are quite distinct and in history philosophy in particular they are recognizably distinct. In other words, this is not an artificial distinction to draw for the history philosophy. The other context it may be that artificial in history of philosophy in terms of whether one is agent-oriented with regard to knowing like the idealist, like the rationalist people of that kind or you are object-oriented, which is much more empiricist or much whatever realist in a Thomistic sense.

7. What is Fuller's current relation to STS? Is it peaceful coexistence or is it something else?

Fuller—At one level let us put it this way. It is evolved or devolved depending with you think it is getting better or worse. I think it is quite clear the earlier period it was much better. But I also saw STS in a very particular way namely as I talk about it especially the *Philosophy and Technology Studies* (2006) as a second moment in a dialectic that has corrected a lot of excesses of normative philosophers who were not sufficiently sensitive to the empirical variety and the historical aspects of knowledge production. So in that respect, the norms of philosophers were putting forward were not really accountable to the way in which knowledge is actually produced. Social epistemology is a way to recast the normative philosophical project in light of STS where STS was playing this dialectical foil of showing all the ways in which the philosophers were wrong. I think that if you go back to the 70s and 80s, a lot of STS was very directly aimed against philosophers and the difference, is it actually had an effect on philosophers. I think with the people of my generation and younger, the tendency has been for philosophy of science to give up on the normative project and do this underlaboring work, conceptual clarification, all that work which is very much in line with STS friendly sort of philosophy of science. I think the problem arises when someone like me who has never stop doubting the normative mission of philosophy and simply want to reconstitute it light of STS but did not want to give it up. I don't think there is as much a place for someone like that since STS is very anti-normative at the end of the day and descriptive. So I think that is been increasing difficult, and STS has become normalized as this descriptive discipline but not normalized so much in the sense of a proper academic discipline. I still think that STS is a little shaky but what is less shaky I think is the idea that that people find STS useful in policy settings to analyze things. STS thrives in a certain way of ignoring these normative questions and get on perfectly without them so that there are many people like me. And so I do think in that respect a certain promise that STS had when I was at Virginia Tech as the peak period for me resulting in *Philosophy, Rhetoric, and End of Knowledge* (2004) because the book that really try to nail it. I think that this started to disappear very soon after. I find STS a kind ever a failed opportunity in a way. But so I do not even think STS has very strong ambitions to become a traditional academic discipline located in the university. It could be free-flowing policy of science and technology, a bit like statistics.

8. What are transhumanities?

Fuller—Transhumanities in terms of the way I use it has to do with reinventing the normative character of the humanities as a person building activity by taking advanced technologies that we have in biology, informatics this stuff as part constituting that a person. So you know if you go back to the old days, the liberal arts curriculum included skills that for the most part did not involve any technology except maybe a musical instrument. But now to train the mind, we require various kinds of technologies to enhance the mind and so forth. These things cannot be just seen as the functioning tools that you take off the shelf but rather they have to be seen as part of a répertoire for a new sense of the integrated self because one of the problems of transhumanism is this tendency to focus on our own particular functionalities that are to be extended. You want to improve your memory, you want to improve your alertness, you want to improve your strength, and these are seen as individual items to be dealt with individually, and it is not clear that you can get an integrated person out of this.

That is why transhumanities is needed. It would be a way to organize what is the proper place for all these different skills and things so that people can have very much like what the classical humanities were.

FR—How would digital humanities would fit into this?

Fuller—Digital humanities is a way of thinking about this. Digital humanities people have a tendency to think that in a sense the traditional humanities will eventually find its way to digital platform. It is a reductionist move, where we will have humanities and that will all be digitized. If once we were able to master the right assemblage of digital skills, we will be able to access all the missed knowledge. I mean there is some interesting stuff that in the spirit so, for example, having to do with curation to get all this data, open data, big data, all this data floating around. How do you exactly organize in a way that can be retrieved properly and be used in a way that actually enhances people in ways that are desirable and so forth? It is greater there is a whole kind of normativity of curation; the digital humanities opens up very nicely, but obviously it is a reductionist move at the end of the day. It is an important part of the whole story but is certainly a very important part in the story certainly to take it seriously.

FR—One last question is that Karl Popper wrote in the *Objective Knowledge* (1994) that knowledge is a searchlight. Would you characterize your view to be similar or is it something different?

Fuller—It is a searchlight? I know where this idea is coming from, and it was very powerful idea in the Enlightenment. I actually think that we are moving to a time where given the amount knowledge that are available that really we are talking more about an economy of epistemic efficiency. That notion has to be very strongly tied with the desire to pursue certain kinds of goals so the ends of knowledge have to be as it were negotiated in tandem with the means of knowledge. I certainly do not believe it is an endless plethora to have knowledge as a kind inherently good thing that will somehow just manages itself and people find their way around it. I do not think that is sufficient. There is going to be a sense the economics of knowledge to have known very strongly implicated. Of course a lot of thinkers have gone along these lines but obviously in a context quite a different from the one we are facing now with data explosion and information access. If you were to look at the people who were talking about it economy of effort with regard to knowledge in the late nineteenth century Mach, James, and Peirce. There is something about this idea, and in our own time Nicholas Rescher is really the man who has come closest to kind to try to get a grip on this is in a philosophically interesting way and of course he gets no credit for this. I credit him but otherwise people do not take him seriously. He is the person who started it and so, for example, the entry on social epistemology in the Encyclopedia of Library and Information Sciences, I spent time on Rescher because I do think he is the person who has got that idea. It is an economy issue and that would be the metaphor I would be working with.

References

Darden, L. (2005). Relations Among Fields: Mendelian, Cytological and Molecular Mechanisms. *Studies in History and Philosophy of Science Part C: Studies in History and Philosophy of Biological and Biomedical Sciences, 36*(2), 349–371.

Fuller, S. (1988). *Social Epistemology*. Bloomington: Indiana University Press.

Fuller, S. (2002). *Knowledge Management Foundations*. Boston: Butterworth-Heinemann.ß.

Fuller, S. (2006). *The Philosophy of Science and Technology Studies*. New York: Routledge.

Fuller, S. (2009). *The Sociology of Intellectual Life*. London: Sage Publications.

Fuller, S. (2014a). *Knowledge: The Philosophical Quest in History*. New York: Routledge.

Fuller, S. (2014b). Social Epistemology: The Future of an Unfulfilled Promise. *Social Epistemology Review and Reply Collective, 3*(7), 29–37. http://wp.me/p1Bfg0-1wG

Fuller, S. (2014c). Keynote Lecture: Russian 'Cosmism' as a Potential Inspiration for Twenty-First Century Interdisciplinary Work. https://www.dur.ac.uk/russianinterdisciplinarity/abstracts/keynote/

Fuller, S., & Collier, J. (2004). *Philosophy, Rhetoric, and the End of Knowledge: A New Beginning for Science and Technology Studies*. New York: Routledge.

Popper, K. (1994). *Objective Knowledge*. Oxford: Oxford University Press.

Remedios, F., & Dusek, V. (2018). *Knowing Humanity in the Social World: The Path of Steve Fuller's Social Epistemology*. New York/Basingstoke: Palgrave Macmillan.

Rouse, J. (2003). Remedios and Fuller on Normativity and Science. *Philosophy of the Social Sciences, 33*, 464–471.

Glossary

Abrahamic religion: Fuller's preferred term for religions based on the Old Testament of the Bible Judaism, Christianity and Islam, where Abraham was the patriarch of the Hebrews

Actor Network Theory (ANT): Theory of society presented by Michael Callon, **Bruno Latour**, and John Law treating society as a network of relations without a dominant center and with no social forces.

Agent oriented epistemology: Focuses on the agent, unifying knowledge in terms of the agent's wordview and purposes. It rejects or the division of intellectual labor and total deference to experts, at most critically accepting expert opinion. It has affinities with **idealism**.

Aquinas, Saint Thomas (1225–1274): Scholastic philosopher of the Dominican order. Aquinas synthesized the Greek, pagan science of Aristotle with doctrines of Christianity. God's attributes, such as goodness, wisdom, and power, are only analogous to human attributes. They are qualitatively different. In the late nineteenth century, Aquinas' philosophy

F.X. Remedios, V. Dusek, *Knowing Humanity in the Social World*,
https://doi.org/10.1057/978-1-137-37490-5

was designated the official philosophy of the Roman Catholic Church.

Arius (256–336): Held doctrine of predestination those saved and those damned from creation independent of individual effort. Arius also held doctrine that Christ did not exist with the Father God from eternity but was created. His doctrine rejected by Catholic Church as heresy.

Behaviorism: Psychological doctrine that science of humans should study explicit behavior only and not appeal to externally unobservable mental events and entities.

Bonaventura, Saint (1221–1274): Franciscan theologian who presents learning as an ascent toward God.

Burke, Edmund (1729–1797): British political thinker and philosopher. A major opponent of the French Revolution. Appealed to by a number of contemporary conservatives.

Carnap, Rudolf: **Logical positivist**: Worked on logic, formal semantics, formalization of science, induction and probability. Early used logic to constructed physical objects from sense experiences.

Classical Sociological Theory: Major sociologists of the early twentieth century (Max Weber, Emile Durkheim Vilfredo Pareto, Georg Simmel, Karl Mannheim), offering comprehensive theories of society.

Converging technologies: The combination of biotechnology, nanotechnology and artificial intelligence.

Cosmism: Doctrine originated by **Fedorov** in late nineteenth century Russia, advocating human resurrection and ultimately human control of the universe. Inspiration for Russian and German space programs as well as the **Vernadsky**'s (See below) notion of the biosphere, and **Teilhard de Chardin**'s noosphere, as well as movements in Russian art and music.

Critical Rationalism: **Karl Popper**'s general approach to knowledge. Involves openness to criticism and willingness to change theses in light of criticism.

Deep Ecology: View that humans should be considered merely a small part of the natural world and that considerations of ecosystems, including animals, plants and physical environment should take precedent over human needs or demands.

Determinism: Doctrine that every event has a cause. In its strongest (Laplacian) form the claim that in principle given the initial positions and velocities of every particle in the universe, all future events can be predicted.

Duns Scotus (1266–1308): Late medieval scholastic philosopher of the Franciscan Order. Scotus defended the thesis that divine attributes were infinite extensions of human attributes, the divine differing from the human only in degree, not in kind. Franciscan philosophers were closer to Plato than to Aristotle.

Ecology: Can refer to the political movement supporting the precautionary principle with respect of environmental damage; but also refers to the branch of biology studying communities of organisms, today involving mathematical models.

Epicurus: (341–270 BCE) Ancient Greek philosopher who held that the universe was made solely of atoms in empty space, the good is simply pleasure, and the gods are indifferent to humans.

Eugenics: Programs and activities to improve the genetics of human populations by breeding superior people (positive eugenics) and/or by eliminating the inferior or unfit people (negative eugenics). Government enforcement of eugenics in the early twentieth century has been replaced by 'free market eugenics" where individuals are able to select qualities of offspring. Fuller advocates eugenics.

Fedorov, Nikolai (1829–1903): Russian founder of Cosmism who applied Russian Orthodox doctrines such as immortality in a secular manner by resurrecting the dead though physical means and collecting the atoms of deceased in outer space. Also envisaged human control of the universe. (See **Tsiolkovksy** and **Vernadsky**).

Gnosticism: Christian heresy holding that the creator of the world was a evil creature, and that God was beyond this realm. Knowledge (*gnosis*) one can move beyond this world to a higher form of being. **Eric Voegelin** (controversially) identified all the modern scientific and technological ambition and utopian political thought as Gnosticism.

Goethe, Johann Wolfgang von (1749–1832) German poet, novelist, scientist. Goethe not only was one of the world's greatest writers, did valuable work in anatomy, geology. Botany, and color theory. A universal man.

Hedgenetics: Individuals form collectives to invest in genetic research and gain access to results of that research. Eugenics would be built into intellectual property law.

Hegel, Georg Wilhem: German philosopher of **Idealism**. Claimed to unify the all of past philosophy and knowledge in a comprehensive system. Presented a scheme of the evolution of spirit accounting for both political and cultural history.

Humboldt, Wilhelm von (1767–1835): Scholar of linguistics and founder of the University of Berlin, model of the modern research university.

Hume, David (1711–1776): Scottish philosopher. Politically conservative traditionalist but also anti-theism. At a rational level we cannot justify causality, induction, external objects or the self. We do attempt to predict the future, generalize from past experience and much else not by reason, but by "custom and habit."

Hutchinson, G. Evellyn (1903–1991): British/American founder of mathematical ecology. Introduced **Vernadsky**'s biosphere and biogeochemistry to US.

Idealism: The doctrine that reality is mental. There are many varieties of idealism. Transcendental Idealism of **Kant, Schelling, and Hegel** dominated German philosophy from the late eighteenth century through the first half of the nineteenth century and dominated British philosophy at the end of the nineteenth century, when it was replaced by analytic philosophy. Idealism emphasizes the creative, constructive powers of the human mind.

Intelligent Design (ID): Theory that living things are the products of an Intelligent Designer. This is presented in opposition to evolution. ID accepts the scientific age of the earth (old earth). Most ID theorists do not explicitly identify the Designer with God primarily for educational law reasons. Fuller does wish to openly identify the Designer as the God of the Abrahamic religions. Fuller associates ID with molecular biology against the natural history approach he attributes to Darwinism.

Interfield Theory: Theory that links theories from two other fields. The relation is horizontal rather than vertical, top down, as in reductionism.

Kant, Immanuel (1724–1804): German idealist philosopher whose First Critique claimed that mind structures and categorizes knowledge, and whose Second Critique bases ethics ones' maxim (ethical guiding principle) being universalized to apply to all other humans.

Karmic religions: Used by Fuller to characterize Asian philosophies that believe in reincarnation, but, more importantly, deny a qualitative difference between humans and non-humans.

Kuhn, Thomas: Historian of science. Author of *The Structure of Scientific Revolutions*, on which most **Science and Technology Studies** is based. Kuhn uses the ambiguous term "paradigm" to characterize worldviews of epochs of sciences and the social framework of a scientific discipline. Kuhn accepts science as it is, refusing to criticize it, and avoids dealing with post-WWII Big Science. Fuller sees Kuhn's views as an artifact of the Cold War.

Kulturwissenschaft: German term for study of culture. Called "science," unlike Anglo-American term.

Kurzweil, Ray: Computer scientists and leading figure in posthumanism. In *The Singularity Is Near* Kurzweil predicts that soon computers will develop intelligence superior to that of present humans.

Lakatos, Imre: Philosopher of science. Student of Popper who attempted to integrate Kuhn's model of continuous normal science with Popper's method of conjectures and refutations and discontinuity of science.

Latour, Bruno: French **STS** thinker who co-authored one of the first anthropological studies of a scientific laboratory. In later work a proponent of **actor network theory**. Latour opposes the critical approach to science and technology studies.

Logical empiricism: a weakened form of **logical positivism** in which sentence need only to be empirically confirmed (partially supported by evidence) rather than conclusively verified. This position resulted from criticisms of the verification principle. Hans Reichenbach and Carl Hempel of the Berlin School were leading logical empiricists.

Logical positivism: Movement from the 1920s to 1960 that claimed science is the only way of knowing and that statements that could not be verified by observation are meaningless. Moritz Schlick, **Rudolf Carnap**, and Otto Neurath were central members of the Vienna Circle of logical positivists.

Mach, Ernst (1838–1916): physicist, psychologist, historian of science around the turn of the twentieth century. Advocated a unity of physics and psychology as alternative orderings of the same data. His unification was not reduction to atoms but a common language. Was the source of later logical positivism.

Merton, Robert King (1910–2003): Sociologist. Described ideals of science: organized skepticism, communalism (sharing of data), universalism, and disinterestedness.

Molecular biology: Biology focusing on DNA and proteins, combining genetics with biochemistry. Fuller contrasts this with **Darwin**'s evolution by natural selection, which he links to natural history, claiming the latter is less scientific.

MOOC: Massive Open Online Course: very large courses on internet, often free.

Naess, Arne (1912–2009): Norwegian analytic philosopher, later the founder of **deep ecology**.

Natural selection: **Darwin's** theory that evolution occurs by random appearance of traits and differential survival due to superior adaptation to the environment.

Naturalism: Doctrine that only natural (non-supernatural) entities exist and all explanations must be in terms of natural phenomena.

Neo-Darwinism: Darwinian **natural selection** complemented by twentieth century population genetics.

Neo-Kantianism: A late nineteenth century movement reviving the philosophy of Immanuel **Kant** (1724–1804) Neo Kantians focused on the theory of knowledge and claimed that different subject matters required different ways of knowing, for instance in the natural versus the social sciences.

Neo-liberalism: Late twentieth century doctrine promulgating market economic treatment of most realms of existence, rejecting and replacing the welfare state, and making the university a purely commercial enterprise.

Object oriented epistemology: Focuses on the objects of knowledge rather than the knower. It accepts separate domains of knowledge, the division of intellectual labor, and defers to experts.

Paradigm: **Thomas Kuhn**'s name for a framework for a field of science, including theory, methodology, notion of ideal theory, possibly metaphysics of fundamental entities. Later distinguished paradigm as exemplar and paradigm as disciplinary matrix.

Pelagius: (354–420) Irish monk who claimed salvation can be attained by human effort, not just grace. Denied predestination and held to free will.

Phenomenology: Approach to philosophy most associated with Edmund Husserl. Involves presuppositionless description of pure experience. Contrasts with **behaviorism** as well as with analytical philosophy of language.

Phlogiston: A substance claimed in the 1700s to exist but now known not to exist. It was released in some chemical reactions and could have negative weight.

Plebscience: Fuller's term for the default form of university science in which teaching is merely an adjunct to scholarship. The analogy is to political plebiscites in which assent if given to already formulated policies.

Polanyi, Michael: Hungarian/British chemist turned social theorist and then philosopher of science. sPolanyi was initiator of **tacit knowledge**. He advocated fealty to prestigious scientists.

Popper, Karl (1902–1994): Austro-British philosopher of science who held that science does not proceed by cautious extrapolations but by bold conjectures and their refutations, the critical approach. Popper's followers and called Popperians.

Posthumanism: Similar to **transhumanism**, but claims humans through technology will be transformed into something that is no longer human.

Postmodernism: Movement of the late twentieth and twenty-first century rejecting comprehensive philosophical systems, unified scientific theories, grand narrative of historical progress, and a unified self.

Postpositivism: Designates a diverse group of philosophers who starting in 1958 rejected the formal account of theories of the logical positivists, using examples from the history of science and accounts from perceptual psychology, emphasizing that empirical observations are structured by the theory held by the observer. Thomas Kuhn, Norwood Russell Hanson, Paul Feyerabend, and Stephen Toulmin are major participants.

Precautionary Principle: Used by ecological activists as well as analysts of technological risk evaluation claiming that if there is any threat of danger from a project it should not be pursued.

Proactionary Principle: Principle that daring and possibly risk posing projects with significant benefits ought to be pursued, opposite of **precautionary principle**.

Prolscience: Fuller's term for a science in which research is evaluated in terms of its service to teaching, in which distribution is integrated into production of knowledge. The term evokes Marx's proletarian of industrial workers seizing the means of production.

Protscience: Fuller's term for science that is democratized. The analogy is to Protestantism as opposed to Roman Catholicism. In Protestantism the believer could read and interpret the Bible for herself, while in Catholicism the church and the priesthood have sole prerogative to interpret the Bible.

Racial Hygiene: Movement begun in early twentieth century Europe empathizing improvement of the race by public health programs and eugenics. Term rejected after WWII because of Nazi use of term included the holocaust.

Reductionism: Notion that scientific theories can be reduced to a more unified theory. One version involves deduction from theories and terms of a more fundamental theory. Another version (which **Mach** and Fuller advocate) is creation of a unified language of science. Means toward the **Unity of Science**.

Satisficing: Economic or decision strategy that pursues a useful of good enough result rather than an optimal result. Theory developed by **Herbert Simon**.

Schelling, Friedrich Wilhelm Joseph (1775–1854): Irrationalist and in later years ultra-conservative romantic idealist philosopher. Claimed that philosophical intuition rather than discursive reasoning and empiricism could give the genius insight into the real nature of things and replace empirical science.

Science and Technology Studies (STS): Studies the topic using sociology, anthropology, rhetorical and literary studies, and cultural studies. Most STS bases itself on **Thomas Kuhn**. STS often opposes itself to philosophy of science as too abstract and non-causal. Fuller criticizes the absence of a normative approach in recent STS.

Simon, Herbert: Inventor of **satisficing**, pioneer of theorem proving and leaning computer programmes.

Singer, Peter (1946–): Australian philosopher who emphasizes the lack of significant difference between humans and other animals. He advocates the "Darwinian Left," an allegedly leftist movement based on Darwin rather than Marx. In contrast of Marxism, Singer emphasizes the biological limits on humans' social aspirations. Singer also advocates "animal liberation" granting rights to non-human animals while advocating abortion as well as infanticide of babies with disabilities or hereditary diseases.

Social Constructivism: A position that claims that social relations as well as scientific theories are constructed by humans. Various pursuers of Science and Technology Studies claim, uncontroversially, that technological equipment and scientific theories are constructed by humans, and, much more controversially, facts and reality are socially constructed.

Social Epistemology: A position rejecting the traditional theories of knowledge that held knowledge to be solely the possession of individuals. Social epistemologists hold that knowledge is gained through socially structured groups. Individualist version explains knowledge in terms of relations similar to those of market economics, while Fuller's version treats the relevant social groups are systems, not merely ties between individuals.

Sociology of Scientific Knowledge (SSK): Sociological view that the content of scientific knowledge can be accounted for by the social structure of science rather than by correspondence with nature.

Synthetic Theory of Evolution: The combination of Darwinian natural selection with Mendelian genetics. Population genetics applied to various fields of biology (such as classification, anatomy, embryology, fossils) in the 1940s and 1950s. Proponents of this theory included Theodosius Dobzahnsky, George Gaylord Simpson, Ernst Mayr.

Tacit knowledge: A kind of non-conscious background knowledge that structures out explicit knowledge. Term from **Michael Polanyi**.

Teilhard de Chardin, Pierre (1881–1955): Catholic priest. Discoverer of Peking Man. Theistic evolutionist claiming convergent evolution and ultimate unity with Christ. Used idea of noosphere of **Vernadsky** as unitary consciousness of humanity.

Theoretical terms: Words or symbols in a theory that are not directly tied to observation and denote entities that are not perceptually observable, or else function as instruments for prediction from which descriptions of observables are deduced. Fuller claims these are supernatural, beyond **naturalism**.

Theosis: Eastern Orthodox doctrine that we can a least partially become one with God.

Transdisciplinarity: Researchers from different fields collaborate to create new theories and methods not limited to the academic disciplines.

Transhumanism: Movement that claims humans can modify their attributes and abilities to have powers that go beyond those of present humans.

Trust: Trust in science is based on belief in authority of experts. The notion is also used by many in science studies and political theory claiming that the basis of social relationships, including those of members of the scientist community, involving acceptance of results of scientists in other fields while lacking specific knowledge of the results. Fuller characterizes as a kind of "**phlogiston**."

Tsiolkovsky, Konstantin (1857–1945): Disciple of **Fedorov** following latter's plan to collect atoms of deceased in outer space, developed some of the first rockets and much of the theory of rocketry, influencing USSR and German rocket programs.

Underlaborer: John Locke (1632–1704) considered the founder of British Empiricism presented his work as merely clearing the rubbish from the edifice of Newton's Physics. Fuller claims that contemporary philosophers of science who analyze particular theories are functioning as underlaborers, not evaluators of the role of science in society.

Unity of Science or Unified science: An description and ideal of science as consisting of a single deductive system with general principles entailing the results of the various scientists or else a single descriptive language unifying the fields. Usually connected with reductionism.

Vernadsky, Vladimir (1863–1945): Follower of Federov. (See above.) Russian founder of geochemistry. First used notion of biosphere, totality of life on earth, and the **noosphere**, totality of consciousness on earth taken up by **Teilhard de Chardin**.

Whorf, Benjamin Lee (1897–1941): Student of Native American languages. Claimed language structures thought; linguistic relativism.

Woefflin, Heinrich (1864–1945): Swiss Art historian. Developer of a formal theory of visual art based on a series of contrasts of formal characteristics.

References

Abir-Am, P. (1983). The Discourse of Physical Power and Biological Knowledge in the 1930s: A Reappraisal of the Rockefeller Foundation's 'Policy' in Molecular Biology. *Social Studies of Science, 12*(3), 241–282.

Aczel, A. D. (2007). *The Jesuit and the Skull: Teilhard de Chardin, Evolution, and the Search for Peking Man*. London: Penquin.

Albrittan, C. C. (1980). *The Abyss of Time: Unraveling the Mystery of the Earth's Age*. San Francisco: Freeman Cooper.

Anderson, P. (1976). *Considerations on Western Marxism*. London: Verso Books.

Anderson, R. (2012, May). Holy Cosmos. *The Atlantic*, No. 4. [online]. Available: www.theatlantic.com/technology/archive/2012/03/the-holy-cosmos-the.../255136. Accessed 30 Jan 2017.

Anon. (1997, December 4). You Can't Follow the Science Wars Without a Map. *The Economist*. http://www.economist.com/node/109188

Arendt, H. (1951). *The Origins of Totalitarianism*. New York: Harcourt Brace.

Arendt, H. (1963). *Eichmann in Jerusalem: A Study in the Banality of Evil*. New York: Viking Press.

Armstrong, R. (2016). *Stararkː A Living, Self Sustaining Spaceship*. New York: Springer Verlag.

Aschoff, J. (1960). Exogenous and Endogenous Components in Circadian Rhythms. *Cold Spring Harbor Symposium Quantitative Biology, 25*, 11–28.

Baker, E., & Oreskes, N. (2017). It's No Game: Post-truth and the Obligations of Science Studies. *Social Epistemology Review and Reply Collective, 6*(8), 1–10. http://wp.me/p1Bfg0-3FB

Barron, C. (2003). A Strong Distinction Between Humans and Non-humans Is No Longer Required for Research Purposes: A Debate Between Bruno Latour and Steve Fuller. *History of the Human Sciences, 16*(2), 77–99.

F.X. Remedios, V. Dusek, *Knowing Humanity in the Social World*, https://doi.org/10.1057/978-1-137-37490-5

Bechtel, W. (Ed.). (1986). *Integrating Scientific Disciplines.* Boston: Kluwer.
Bechtel, W. (1993). Integrating Sciences by Creating New Disciplines: The Case of Cell Biology. *Biology and Philosophy, 8*, 277–299.
Bechtel, W., & Abrahamson, A. (2009). Decomposing, Recomposing, and Situating Circadian Mechanisms: Three Tasks of Developing Mechanistic Explanations. In *Publications of the Austrian Wittgenstein Society: New Series* (Vol. 12). Frankfurt: Ontos Verlag.
Behe, M. (1996). *Darwin's Black Box: The Biochemical Challenge to Evolution.* Glencoe: The Free Press.
Behe, M. (2007). *The Edge of Evolution: The Search for the Limits of Darwinism.* Glencoe: The Free Press.
Bell, R. (1992). *Impure Science: Fraud, Compromise, and Political Influence in Scientific Research.* New York: Wiley.
Berger, R. L. (1990). Nazi Science: The Dachau Hypothermia Experiments. *New England Journal of Medicine, 322*(20), 1435–1440.
Block, N., & Dworkin, G. (1976). *The IQ Controversy.* New York: Pantheon.
Boltzmann, L. (1974). In B. McGuiness (Ed.), *Theoretical Physics and Philosophical Problems.* Boston: D. Reidel Publishing Company.
Bonaventura, S. (1952). *The Mind's Road to God.* Indianapolis: Liberal Arts Press.
Bostrom, N. (2014). *Superintelligence.* Oxford: Oxford University Press.
Boycott, A. S., Damant, G. C. C., & Haldane, J. S. (1908). The Prevention of Compressed Air Illness. *Journal of Hygiene, 8*(3), 342–443.
Brown, A. (2006, March 6). When Evolutionists Attack. *The Guardian.* http://www.theguardian.com/world/2006/mar/06/religion.uk
Brown, F. A., Jr. (1960). Response to Pervasive Geophysical Factors and the Biological Clock Problem. *Cold Spring Harbor Symposia in Quantitative Biology, 25*, 57–71.
Brown, F. A., Woodland Hastings, J., & Palmer, J. D. (1970). *The Biological Clock: Two Views.* New York: Academic Press.
Callon, M. (1986). Some Elements of a Sociology of Translation: Domestication of the Scallops and the Fishermen of St Brieuc Bay. In J. Law (Ed.), *Power, Action and Belief: A New Sociology of Knowledge* (pp. 196–233). London: Routledge & Kegan Paul.
Cambrosio, A., & Keating, P. (1983). The Disciplinary Stake: The Case of Chronobiology. *Social Studies of Science, 13*(3), 323–353.
Caplan, A. (Ed.). (1992). *When Medicine Went Wrong: Bioethics and the Holocaust.* Totowa: Humana Press.
Cartwright, N. (1983). *How the Laws of Physics Lie.* Oxford: Oxford University Press.
Cassirer, E. (1906–1907, 1910). *Das Erkenntnisproblem in der Philosophie und Wissenschaft der Neueren Zeit* (Vols. 1–3). Berlin: Bruno Cassirer Verlag.
Cassirer, E. (1944). *The Problem of Knowledge.* New Haven: Yale University Press.

Cassirer, E. (1951). *The Philosophy of the Enlightenment.* Princeton: Princeton University Press.
Chamovitz, D. (2013). *What a Plant Knows: A Guide to the Sense.* New York: Farrar, Straus, and Giroux.
Chargaff, E. (1980). "Gullible's Travels," and "Matches for Hersostratus". In *Heraclitean Fire* (pp. 100–103). New York: Warner Books.
Chase, A. (2003). *Harvard and the Unibomber: The Education of an American Terrorist.* New York: Norton.
Clifford, W. K. (1876, December –1877, May). The Ethics of Belief. *Contemporary Review, 29,* 288–309.
Coeckelburgh, M. (2013). *Human Beings @ Risk.* Dordrecht: Springer.
Collin, F. (2013). Two Kinds of Social Epistemology. *Social Epistemology Review and Reply Collective, 2*(8), 79–104. http://wp.me/p1Bfg0-Ul
Collins, A. (1998). *In the Sleep Room: The Story of CIA Brain Washing Experiments in Canada.* Toronto: Key Porter Books.
Collins, H. (2007). *Rethinking Expertise.* Chicago: University of Chicago Press.
Collins, H. (2010). *Tacit and Explicit Knowledge.* Chicago: University of Chicago Press.
Coolidge, J. L. (1950). *The Mathematics of Great Amateurs.* Oxford: The Clarendon Press.
Corrigan, O. (2003). Empty Ethics: The Problem of Informed Consent. *Sociology of Health and Illness, 25*(7), 768–792.
Cozzi, N. V. (2017, Spring). Psychedelic Breakthroughs in Neuroscience: How Psychedelic Drugs Influenced the Growth of Neuroscience. *MAPS Bulletin.*
Daan, S. (1998, May 9). *Colin Pittendrigh, Jürgen Aschoff, and the Natural Entrainment of Circadian Systems.* In Presented at the 6th Meeting of the Society for Research on Biological Rhythms, Amelia Island, np.
Daan, S. (2010). A History of Chronobiological Concepts. In U. Albrecht (Ed.), *The Circadian Clock* (Vol. 12, pp. 1–35). New York: Springer.
Darden, L. (2005). Relations Among Fields: Mendelian, Cytological and Molecular Mechanisms. *Studies in History and Philosophy of Science Part C: Studies in History and Philosophy of Biological and Biomedical Sciences, 36*(2), 349–371.
Darden, L., & Maull, N. (1977). Interfield Theories. *Philosophy of Science, 44,* 43–64.
Darwin, C. (2015). *The Formation of Vegetable Mould Through the Action of Worms, with Observation of Their Habits.* New York: Dossier Press. (Orig. 1882).
de Price, D. S. (1967). Gods in Black Boxes. In E. Bowles (Ed.) *Computers and Humanistic Research.* Englewood: Prentice Hall. Quoted in Heims, S. (1980). *John von Neumann and Norbert Wiener.* Cambridge, MA: The MIT Press.
Deevey, E. (1960). The Hare and the Haruspex: A Cautionary Tale. *American Scientist, 48*(3), 415–430.

Delgado, J. M. (1969). *Physical Control of the Mind: Toward a Psychocivilized Society*. New York: Harper.

Downes, S. (2000). Review of *Science* by Steve Fuller. *Philosophy of the Social Sciences, 30*(1), 140–145.

Downes, S. (2001). Agents and Norms in a New Economics of Science. *Philosophy of the Social Sciences, 31*, 224–238.

Dusek, V. (1966a). Continental Continuum Theory. https://www.researchgate.net/publication/277719895_Continental_Continuum_Theory

Dusek, V. (1966b). The Hamiltonian Revolution in Physics. https://www.researchgate.net/publication/273774695_The_Hamiltonian_Revolution_in_Physics

Dusek, V. (1990). The Bukharin Delegation on Science and Society: Action and Reaction in British Science Studies. In N. N. Kozlov & E. D. Weiss (Eds.), *Nikolai Ivanovich Bukharin: A Centenary Appreciation* (pp. 129–147). New York: Praeger Publishers.

Dusek, V. (2000). Aristotle's Four Causes and Contemporary 'Newtonian' Dynamics. In D. Sfendoni-Mentzou, J. N. Hattagiani, & D. Johnston (Eds.), *Aristotle and Contemporary Science* (pp. 81–93). New York: Peter Lang Publishers.

Dusek, V. (2014). [Review of] Bruno Latour, *An Inquiry into the Modes of Existence. Notre Dame Philosophical Review*. http://ndpr.nd.edu/news/an-inquiry-into-modes-of-existence-an-anthropology-of-the-moderns/

Edgerton, S. Y. (1991). *The Heritage of Giotto's Geometry: Art and Science on the Eve of the Scientific Revolution*. Ithaca: Cornell University Press.

Edmunds, L. N., Jr. (2012). *Cellular and Molecular Bases of Biological Clocks: Models and Mechanisms for Circadian Timekeeping*. New York: Springer.

Ehrlich, P. R. (1968). *The Population Bomb*. New York: Ballantine Books.

Elgin, C. (1999). *Considered Judgment*. Princeton: Princeton University Press.

Elgin, C. (2013). Epistemic Agency. *Theory and Research in Education, 11*(2), 135–152.

Elliott, C. (2014, September 3). More or Less than Human? *New Scientist*.

Elster, J. (1979). *Logic and Society*. Chichester: Wiley.

Engber, D. (2016, January 26). The Neurologist Who Hacked His Brain—And Almost Lost His Mind. *Wired*.

Epstein, S. (1998). *Impure Science: AIDS, Activism, and the Politics of Knowledge*. Berkeley: University of California Press.

Eschner, K. (2016, December). The Story of the Real Canary in the Coal Mine. *Smithsonian Magazine*. http://www.smithsonianmag.com/smart-news/story-real-canary-coal-mine-180961570/

Fedorov, N. F. (1990). *What Was Man Created For? The Philosophy of the Com Mon Task*. Bath: Honeyglen Publishing.

Floridi, L. (2002). On the Intrinsic Value of Information Objects and the Infosphere. *Ethics and Information Technology, 4*(4), 287–304.

Foster, R. G., & Kretzman, L. (2005). *Rhythms of Life: The Biological Clocks That Control the Daily Lives of Everything*. New Haven: Yale University Press.
Friel, H. (2010). *The Lomborg Deception: Setting the Record Straight About Global Warming*. New Haven: Yale University Press.
Frodeman, R. (2013). *Sustainable Knowledge: A Theory of Interdisciplinarity*. London: Palgrave.
Frodeman, R. (2014). Sustainable Knowledge: An Exchange, Davis, Dieleman, Frodeman, Remedios, Riggio, Simbürger, Suomela. https://social-epistemology.com/2014/03/15/sustainable-knowledge-an-exchange-davis-dieleman-remedios-riggio-simburger-suomela/#more-5189
Frodeman, R. (2015). Anti-Fuller: Transhumanism and the Proactionary Imperative. *Social Epistemology Review and Reply Collective, 4*(4), 38–43. http://wp.me/p1Bfg0-1Zr
Fuller, S. (1985). *Bounded Rationality in Law and Science*. PhD Dissertation, University of Pittsburgh.
Fuller, S. (1988). *Social Epistemology*. Bloomington: Indiana University Press.
Fuller, S. (1992). Being There with Thomas Kuhn: A Parable for Postmodern Times. *History and Theory, 31*(3), 241–275.
Fuller, S. (1993). Social Constructivism Teaching Itself a Lesson: Science Studies as a Social Movement. *Danish Yearbook of Philosophy, 28*, 47–60.
Fuller, S. (1996a). Social Epistemology and the Recovery of the Normative in the Postepistemic Era. *Journal of Mind and Behavior, 17*(2), 83–97.
Fuller, S. (1996b). Social Epistemology and Psychology. In W. O'Donohue & R. Kitchener (Eds.), *The Philosophy of Psychology*. London: Sage Publications.
Fuller, S. (1997a). Making Science an Experimenting Society. In W. Dunn (Ed.), *The Experimenting Society: Essays in Honor of Donald T. Campbell* (pp. 69–102). New Brunswick: Transaction Publishers.
Fuller, S. (1997b). *Science*. Buckingham: Open University Press.
Fuller, S. (1999). *The Governance of Science: Ideology and the Future of the Open Society*. Buckingham: Open University Press.
Fuller, S. (2000). *Thomas Kuhn: A Philosophical History of Our Times*. Chicago: Chicago University Press.
Fuller, S. (2000a, February 22). E-mail to Remedios.
Fuller, S. (2000b). *Thomas Kuhn: A Philosophical History of Our Times*. Chicago: Chicago University Press.
Fuller, S. (2002). *Social Epistemology* (2nd ed.). Bloomington: Indiana University Press.
Fuller, S. (2002a). *Social Epistemology* (2nd ed.). Bloomington: Indiana University Press.
Fuller, S. (2002b). Demysitfying Gnostic Scientism. *Rhetoric & Public Affairs, 5*(1), 133–157.
Fuller, S. (2002c). *Knowledge Management Foundations*. Boston: Butterworth-Heinemann.ß.

Fuller, S. (2004). *Kuhn vs. Popper: The Struggle for the Soul of Science*. New York: Columbia University Press.

Fuller, S., & Collier, J. (2004). *Philosophy, Rhetoric, and the End of Knowledge: A New Beginning for Science and Technology Studies* (2nd ed.). Mahwah: Lawrence Erlbaum Associates, Publisher.

Fuller, S. (2005a). Is Science Studies Revolutionary or Merely Revolting? *Science Studies, 18*(1), 75–83.

Fuller, S. (2005b). *The Intellectual*. London: Icon Books.

Fuller, S. (2006a). Humanity as the Site of Ideological Conflict in the Twenty-First Century. *Ludus Vitalis, 14*(26), 65–88.

Fuller, S. (2006b). *The New Sociological Imagination*. London: Sage Publications, Ltd.

Fuller, S. (2006c). *The Philosophy of Science and Technology Studies*. New York: Routledge.

Fuller, S. (2007a). *New Frontiers in Science and Technology Studies*. New York: Polity Press.

Fuller, S. (2007b). *The Knowledge Book: Key Concepts in Philosophy, Science, and Culture*. Montreal: McGill-Queens University Press.

Fuller, S. (2007c). *Science vs. Religion: Intelligent Design and the Problem of Evolution*. Cambridge: Polity Press.

Fuller, S. (2008a). *Dissent on Descent: Intelligent Design's Challenge to Darwinism*. Cambridge: Icon Books.

Fuller, S. (2008b). Science Studies Goes Public: Report on an Ongoing Performance. *Spontaneous Generations, 2*(1), 11–21.

Fuller, S. (2009a). *The Sociology of Intellectual Life*. London: Sage Publications.

Fuller, S. (2009b). Philosophy of Science in an Age of Neo-Darwinian Apologetics. *Ludus Vitalis, XXVII*(3), 247–257.

Fuller, S. (2010). *Science*. Durham: Acumen Publishing Limited.

Fuller, S. (2011a). *Humanity 2.0: What It Means to Be Human Past, Present and Future*. London: Palgrave Macmillan.

Fuller, S. (2011b). The Posthuman Challenge to Ecological Correctness. https://www.youtube.com/watch?v=r_lzFSnxYZA. https://www.youtube.com/watch?v=UQJ_mWC4dpc

Fuller, S. (2011c). Evidence, What Evidence? *Philosophy of the Social Sciences, 41*(4), 567–573.

Fuller, S. (2012). Social Epistemology: A Quarter Century Itinerary. *Social Epistemology, 26*(3–4), 267–283.

Fuller, S. (2013a). *Preparing for Life in Humanity 2.0*. New York/Basingstoke: Palgrave Macmillan.

Fuller, S. (2013b). The Dawn of Critical Neuroscience. *History of the Hunab Sciences, 26*(3), 107–115.

Fuller, S. (2014a). *Knowledge: The Philosophical Quest in History*. New York: Routledge.

Fuller, S. (2014b). Comments to J. Britt Holbrook 'Fuller's Categorical Imperative: The Will to Proaction'. *Social Epistemology Review and Reply Collective, 2*(11), 20–26. http://wp.me/p1Bfg0-13K

Fuller, S. (2014c). Social Epistemology: The Future of an Unfulfilled Promise. *Social Epistemology Review and Reply Collective, 3*(7), 29–37. http://wp.me/p1Bfg0-1wG

Fuller, S. (2014d). Keynote Lecture: Russian 'Cosmism' as a Potential Inspiration for Twenty-First Century Interdisciplinary Work. https://www.dur.ac.uk/russianinterdisciplinarity/abstracts/keynote/

Fuller, S. (2014e). Towards a Proactionary Welfare State. *Social Epistemology Review and Reply Collective, 3*(5), 82–84. http://wp.me/p1Bfg0-1sx

Fuller, S. (2014f). *From Transfiguration to Transhumanism: Taking Seriously Theology as the Science of God*. ABC Religion and Ethics No. 30, September.

Fuller, S. (2015a). Retrieved January 28, 2017 from https://twitter.com/professorstevefuller/status/660426459547242497

Fuller, S. (2015b, April 4). Knowing Knowledge. *Social Epistemology Review and Reply Collective*. https://social-epistemology.com/2015/04/12/knowing-knowledge-iii-adam-riggio-and-steve-fuller/. Accessed 17 Feb 2017.

Fuller, S. (2015c, September 30). Who's Afraid of the Human? A Theological Defence of Ecomodernism. *ABC Religion and Ethics.*

Fuller, S. (2016). *Science Has Always Been a Bit 'Post-Truth'*. https://www.theguardian.com/science/politicalscience/2016/dec/15/science-has-always-been-a-bit-post-truth

Fuller, S. (2016a). *The Academic Caesar: University Leadership Is Hard*. London: Sage Publications.

Fuller, S. (2016b). A Sense of Epistemic Agency Fit for Social Epistemology. In P. Reider (Ed.), *Social Epistemology and Epistemic Agency* (pp. 21–42). London: Rowman International.

Fuller, S. (2016c). Symposium on Steve Fuller's Knowledge: The Philosophical Quest in History. *Metsacience, 25*(1), 3–23.

Fuller, S. (2016d). *Letter, First Things*, January.

Fuller, S (2017a). The Case for Academic Caesars. *Times Higher Education Magazine*. https://www.timeshighereducation.com/features/case-academic-caesars

Fuller, S. (2017b). Is STS All Talk and No Walk? *EASST Review, 36*(1) https://easst.net/article/is-sts-all-talk-and-no-walk/

Fuller, S. (n.d.). Steve Fuller on Malachi Cohen. http://www.the-rathouse.com/shortreviews/Fuller-on-Hacohen.html

Fuller, S., & Collier, J. (2004). *Philosophy, Rhetoric, and the End of Knowledge: A New Beginning for Science and Technology Studies* (2nd ed.). Mahwah: Lawrence Erlbaum Associates.

Fuller, S., & Collier, J. (2004). *Philosophy, Rhetoric, and the End of Knowledge: A New Beginning for Science and Technology Studies* (2nd ed.). Mahwah: Lawrence Erlbaum Associates, Publisher.

Fuller, S., & Lipinska, V. (2014a). *The Proactionary Imperative: A Foundation for Transhumanism*. New York/Basingstoke: Palgrave Macmillan.

Fuller, S., & Lipinska, V. (2014b). Transhumanism. *Social Epistemology Review and Reply Collective, 3*(11), 25–29.

Fuller, S., Chioti, R., & Ernstsons, K. (2017). Connecting with the Divine and the Sacred and Becoming Cosmically Conscious. In R. Armstrong (Ed.), *Star Ark* (pp. 383–409). New York: Springer Verlag.

Gage, J. (1999). *Color and Culture: Practice and Meaning from Antiquity to Abstraction*. Berkeley: University of California Press.

Gass, W. (1975). *On Being Blue*. New York: New York Review of Books Editions.

Gayon, J. (1998). *Darwin's Struggle for Survival*. Cambridge: Cambridge University Press.

Gibbs, M. L. (2011). *Love Canal: The Story Continues*. Washington: Island Press.

Gillispie, C. C. (1959). *Genesis and Geology: The Impact of Scientific Discoveries Upon Religious Beliefs in the Decades Before Darwin*. Cambridge: Harvard University Press.

Goldberg, S. (2016). A Proposed Research Program for Social Epistemology. In P. J. Reider (Ed.), *Social Epistemology and Epistemic Agency* (pp. 3–20). London: Rowman and Littlefield International.

Goldman, A. (2010). Systems-Oriented Social Epistemology. In T. Gendler & J. Hawthorne (Eds.), *Oxford Studies in Epistemology* (Vol. 3, pp. 189–214). Oxford: Oxford University Press.

Goodwin, B. (1963). *Temporal Rhythms in Cells: A Dynamic Theory of Cellular Control Processes*. Cambridge, MA: Academic Press.

Gorley, M., & Fellet, M. F. (2015). The Pauling-Teller Debate: A Tangle of Expertise and Values. *Issues in Science and Technology, 31*(4, Summer), 78–82.

Gould, S. J. (1983). A Worm for the Century. In *Hen's Teeth and Horse's Toes* (pp. 120–133). New York: Norton.

Gould, S. J., Raup, D. M., Sepkoski, J., & Schopf, J. M. (1977). The Shape of Evolution: a Comparison of Real and Random Clades. *Paleobiology, 3*(1), 23–40.

Graham, L., & Kantor, J. M. (2009). *Naming Infinity: A True Story of Religious Mysticism and Mathematical Creativity*. Cambridge, MA: Harvard University Press.

Gray, J. (2011). *The Immortality Commission: Science and the Strange Quest of Defeat Death*. New York: Farrar, Straus, and Giroux.

Grayling, A. C. (2008, September). Origin of the Specious. *The New Humanist*. https://newhumanist.org.uk/articles/1856/origin-of-the-specious

Greene, J. C. (1961). *The Death of Adam*. New York: New American Library.

Grunbaum, B., & Shephard, G. C. (2016). *Tilings and Patterns* (2nd ed.). New York: Dover.
Guston, D. H. (2002). Secularising Science? *Futures, 34*, 197–199.
Hacking, I. (1983). *Representing and Intervening*. Cambridge: Cambridge University Press.
Hacohen, M. H. (2000). *Karl Popper: The Formative Years*. Cambridge: Cambridge University Press.
Halberg, F., et al. (2003). Transdisciplinary Unifying Implications of Circadian Findings in the 1950s. *Journal of Circdadian Rhythms, 1*(2), np. http://www.jcircadianrhythms.com/articles/10.1186/1740-3391-1-2
Harries, K. (2002). *Infinity and Perspective*. Cambridge, MA: The MIT Press.
Harrison, P. (2016). Science, Eastern Orthodoxy and Protestantism. *Isis, 107*(3), 587–591.
Hinton, P. (2013). *The Mass Observers: A History, 1937–1949*. Oxford: Oxford University Press.
Hobsbawm, E., & Ranger, T. O. (1982). *The Invention of Tradition*. Cambridge: Cambridge University Press.
Hobsbawm, E., & Ranger, T. (1984). *The Invention of Tradition*. Cambridge: Cambridge University Press.
Holbrook, J. B. (2013). Fuller's Categorical Imperative: The Will to Proaction. *Social Epistemology Review and Reply Collective, 2*(11), 20–26. http://wp.me/p1Bfg0-13K
Holbrook, B., & Adam, B. (2013). Knowing and Acting: The Precautionary and Proactionary Principles in Relation to Policy Making. *Social Epistemology Review and Reply Collective, 2*(5), 15–37. http://wp.me/p1Bfg0-KQ
Hull, D. (1988). *Science as Process: An Evolutionary Account of the Social Development of Science*. Chicago: University of Chicago Press.
Humphreys, L. (1975). *Tearoom Trade: Impersonal Sex in Public Places (Observations)*. London: Routledge.
Hutchinson, G. E. (1948a). Circular Causal Systems in Ecology. *Annals of the New York Academy of Sciences, 50*(4), 221–246.
Hutchinson, G. E. (1948b). On Living in the Biosphere. *Science, 108*, 587.
Hutchinson, G. E. (1950). A Survey of Contemporary Knowledge of Biogeochemistry III. *Bulletin of the American Museum of Natural History, 96*, 1–544.
Hutchinson, G. E. (1953). The Gothic Attitude to Natural History. In *The Itinerant Ivory Tower* (pp. 31–35). New Haven: Yale University Press.
Hutchinson, G. E. (1954). *Scientific Language and Religious Faith*. New York: National Council of the Episcopal Church.
Hutchinson, G. E. (1970). The Biosphere. *Scientific American, 223*(3), 45–53.
Ivins, W. M. (1946). *Art and Geometry: A Study of Space Intuitions*. Cambridge, MA: Harvard University Press.

James, J. (1993). *Music of the Spheres: Music, Science, and the Natural Order of the Universe*. New York: Springer-Verlag.

Jeans, J. (1937). *Music and Science*. Cambridge: Cambridge University Press.

Jonas, H. (1967). Judaism, Christianity and the Western Tradition. *Commentary*, (41), 61–70. https://www.commentarymagazine.com/articles/judaism-christianity-the-western-tradition/

Jonas, H. (2001). *The Gnostic Religion* (3rd edn.). Boston, MA: Beacon Press. (Orig. *Gnosis und Satanic Geist*, Vol. 1 1934, Vol. 2 1954).

Kazantzakis, N. (1960). *Saviors of God*. New York: Simon and Schuster.

Kitcher, P. (2007). *Living with Darwin*. Oxford: Oxford University Press.

Kitzmiller, T. (2005). The Middle District of Pennsylvania: Tammy Kitzmiller, et al.: Case No. 04cv2688 Plaintiffs : Judge Jones v. Dover Area School District, *Tammy Kitzmiller, et al.* v. *Dover Area School District, et al.* (400 F. Supp. 2d 707, Docket no. 4cv2688).

Klitzman, R. L. (2015). *Ethics Police: The Struggle to Make Human Research Safe*. Oxford: Oxford University Press.

Kosinski, J. (1970). *Being There*. New York: Harcourt Brace.

Krimsky, S. (2004). *Science in the Private Interest: Has Corrupted Biomedical Research?* Lanham, MD: Rowan and Littlefield.

Kuhn, T. (1977). *The Essential Tension*. Chicago: University of Chicago Press.

Kuhn, T. (1992). *The Trouble with the Historical Philosophy of Science*. Robert and Maurine Rothschild Distinguished Lecture, 19 November 1991. An Occasional Publication of the Department of the History of Science. Cambridge, MA: Harvard University Press.

Kuhn, T. S. (1957). The Function of Dogma in Science. In M. Claggett (Ed.), *Critical Problems in the History of Science* (pp. 347–369). Madison: University of Wisconsin Press.

Kuhn, T. S. (1970). *The Structure of Scientific Revolutions* (2nd ed.). Chicago: University of Chicago Press.

Kuhn, T. S. (1987). *Black Body Theory and the Quantum Discontinuity*. Chicago: The University Press.

Kuhn, T. S., et al. (1967). *Sources for History of Quantum Physics*. Philadelphia: American Philosophical Society.

Kuhn, T., et al. (2000). *The Road Since Structure*. Chicago: The University of Chicago Press.

Kurzweil, R. (2005). *The Singularity Is Near: When Humans Transcend Biology*. New York: Penguin Group.

Labinger, J. (2013). *Up from Generality: How Inorganic Chemistry Finally Became a Respectable Field*. New York: Springer Verlag.

Lakatos, I. (1970). Falsification and the Methodology of Scientific Research Programmes. In I. Lakatos & A. Musgrave (Eds.), *Criticism and the Growth of Knowledge* (pp. 91–196). Cambridge: Cambridge University Press.

Latour, B. (1987). *Science in Action, How to Follow Scientists and Engineers Through Society*. Cambridge, MA: Harvard University Press.
Latour, B. (1993). *We Have Never Been Modern*. Cambridge, MA: Harvard University Press.
Latour, B. (2004a). Why Has Critique Run Out of Steam? *Critical Inquiry, 30*, 225–248.
Latour, B. (2004b). *The Politics of Nature*. Cambridge, MA: Harvard University Press.
Latour, B. (2007). *Reassembling the Social: An Introduction to Actor-Network-Theory*. Oxford: Oxford University Press.
Latour, B. (2013). *An Inquiry into Modes of Existence: An Anthropology of the Moderns* (trans. Porter, C.). Cambridge MA: Harvard University Press.
Lattin, D. (2011). *The Harvard Psychedelic Club*. Boston: Harper.
Lee, M. A., & Shain, B. (1994). *Acid Dreams*. New York: Grove Press.
LeRoy, E. (1918). *What Is Dogma?* Chicago: The Open Court Publishing Company.
Lessl, T. M. (2002). Gnostic Scientism and the Prohibition of Questions. *Rhetoric & Public Affairs, 5*(1), 133–157.
Levenson, T. (1994). *Measure for Measure: A Musical History of Science*. New York: Simon and Schuster.
Levi-Leblond, J.-M. (1975). Ideology of/in Contemporary Physics. In S. Rose & H. Rose (Eds.), *The Radicalization of Science* (pp. 136–175). London: The Macmillan Press.
Levine, A. G. (1982). *Love Canal: Science, Politics, People*. Lanham, MD: Rowan and Littlefield.
Levins, R. (1968). *Evolution in Changing Environments*. Princeton: Princeton University Press.
Lifton, R. J. (1988). *The Nazi Doctors*. New York: Basic Books.
Lloyd, E. A. (1994). *The Structure and Confirmation of Evolutionary Theory*. Princeton: Princeton University Press.
Locke, J. (1894). *An Essay on Human Understanding*. Oxford: Clarendon Press. (Orig. 1689).
Lomborg, B. (2001). *The Skeptical Environmentalist*. Cambridge: Cambridge University Press.
Lomborg, B. (2007). *Cool It: The Skeptical Environmentalist's Guide to Global Warming*. New York: Norton Publishing.
Lowrey, P. L., Menaker, M., et al. (2000). Positional Synthenic Cloning and Functional Characterization of the Mammalian Circadian Mutation tau. *Science, 288*(5465), 484–491.
Lucretius Titus Carus. (2008). *[94 BCE] De Rerum Natura: On the Nature of Things* (trans. Slavitt, D.). Berkeley: University of California Press.

Lynch, W. (2001). *Solomon's Child*. Stanford: Stanford University Press.
Lynch, W. (2016). Social Epistemology Transformed: Steve Fuller's Account of Knowledge as a Divine Spark for Human Domination. *Social Epistemology Review and Reply Collective*. https://social-epistemology.com/2016/05/26/social-epistemology-transformed-steve-fullers-account-of-knowledge-as-a-divine-spark-for-human-domination-william-t-lynch/
Lynch, W. T. (2017). Cultural Evolution and Social Epistemology: A Darwinian Alternative to Steve Fuller's Theodicy of Science. *Social Epistemology, 31*(2), 224–234.
Malapi-Nelson, A. (2016). Transhumanism, Christianity and Modern Science: Some Clarifying Points Regarding Shiffman's Criticism of Fuller. *Social Epistemology Review and Reply Collective, 5*(2), 1–5.
Margulis, L., & Sagan, D. (1997). *Slanted Truths: Essays on Gaia, Symbiosis, and Evolution*. Seacaucus: Copernicus Books.
Mark, V. H., & Frank, R. E. (1970). *Violence in the Brain*. Hagerstown: Harper and Row.
Maull, N. (1977). Unifying Theories Without Reduction. *Studies in History and Philosophy of Science, 8*(2), 143–162.
McDougall, W. A. (1985). *The Heavens and Earth*. New York: Basic Books.
McHale, D. (1985). *George Boole: His Life and Work*. Dublin: Boole Press, Ltd.
McShea, D. W., & Brandon, R. N. (2010). *Biology's First Law*. Chicago: University of Chicago Press.
Meadows, D. H. (1972). *Limits to Growth*. New York: Signet Books.
Medvedev, Z. (1974). Caucasus and Altaic Longevity: A Biological or Social Problem. *The Gerontologist, 14*(5 Part 1), 381–387.
Menaker, M. (1968). Extraretinal Light Perception in Sparrow, I. Entrainment of the Biological Clock. *Proceedings of the National Academy of Sciences USA, 59*(2), 414–421.
Mesthene, I. (1967). Technology and Wisdom. In I. Mesthene (Ed.), *Technology and Social Change* (pp. 109–115). Indianapolis: Bobbs-Merrill.
Milgram, S. (1974). *Obedience to Authority: An Experimental View*. New York: Harper & Row.
Mirowski, P. (2009). *Science-Mart*. Harvard: Harvard University Press.
Mirowski, P., & Sent, E.-M. (2007). The Commercialization of Science, and the Response of STS. In E. J. Hackett, O. Amsterdamska, M. Lynch, & J. Wajcman (Eds.), *The Handbook of Science & Technology Studies* (3rd ed., pp. 635–690). Cambridge, MA: MIT Press.
More, M. (2005). *The Proactionary Principle*. http://www.maxmore.com/proactionary.html
More, M. (2010). The Overhuman in the Transhuman. *Journal of Evolution and Technology, 21*(1), 1–4. http://jetpress.org/more.htm
Mudd, S. (2013). Rethinking the Priority of Practical Reason. *European Journal of Philosophy, 23*, 1–25.

Naess, A. (1938). *'Truth' as Conceived by Those Who Are Not Professional Philosophers.* Oslo: I kommisjon hos J. Dybwad. Skrifter utg. av det Norske videnskaps-akademii Oslo, II. Hist.-filos. Klasse.

Negley, G. (1942). *The Organization of Knowledge.* New York: Prentice Hall.

Neufeld, M. J. (2007). *Von Braun: Dreamer of Space, Engineer of War.* New York: Vintage Books.

Nordhaus, W. D. (2016). Revisiting the Global Cost of Carbon. *PNAS, 114*(7), 1518–1523.

O'Neil, O. (2003). Some Limits of Informed Consent. *Journal of Medical Ethics, 29*(1), 4–7.

Oberg, J. (1984). *The New Race for Space.* Harrisburg: Stackpole Books.

Oronofsky, D. (1998). *Pi,* Protozoa Pictures.

Pagels, E. (1989). *The Gnostic Gospels.* New York: Vintage.

Panofsky, E. (1962). Artist, Scientist, Genius: Notes on the Renaissance-Daemmerung. In W. K. Ferguson et al. (Eds.), *The Renaissance: Six Essays* (pp. 123–182). New York: Harper and Row Publishers.

Parmet, W. E. (1985). AIDS and Quarantine. The Revival of an Archaic Doctrine. *Hofstra Law Review, 14*(1), 11–53.

Pedersen, D. (2013). Who Should Govern the Welfare State 2.0? A Comment on Fuller. *Social Epistemology Review and Reply Collective, 2*(12), 51–59. http://wp.me/p1Bfg0-19F

Peirce, C. S. (1958). Note on the Theory of the Economy of Research. In A. W. Burks (Ed.), *Collected Papers* (Vol. 7, pp. 76–83). Cambridge, MA: Harvard University Press.

Penfield, W. (1959). *Speech and Brain Mechanisms.* Princeton: Princeton University Press.

Penfield, W. (1975). *The Mystery of the Mind.* Princeton: Princeton University Press.

Pérez-Ramos, A. (1988). *Francis Bacon's Idea of Science and the Maker's Knowledge Tradition.* Oxford: Oxford University Press.

Pierce, J. R. (1983). *The Science of Musical Sound.* New York: Freeman.

Pittendrigh, C. S. (1960). Circadian Rhythms and Circadian Organization of Living Systems. *Cold Spring Harbor Symposia in Quantitative Biology, 24,* 159–184.

Polanyi, M. (1958). *Personal Knowledge: Towards a Postcritical Philosophy.* Chicago: Chicago University Press.

Polanyi, M. (1966). *The Tacit Dimension.* Garden City: Anchor Books.

Popper, K. (1994). *Objective Knowledge.* Oxford: Oxford University Press.

Putnam, H. (1970, December 31). Personal Communication with Dusek.

Pyman, A. (2010). *Pavel Florensky: A Quiet Genius.* New York: Continuum International Publishing Group.

Raup, D., & Gould, S. J. (1974). Stochastic Simulation and Evolution: Toward a Nomothetic Paleontology. *Systematic Zoology, 23*(3), 305–322.
Remedios, F. (2003a). *Legitimizing Scientific Knowledge*. Lanham: Lexington Books.
Remedios, F. (2003b). Fuller and Rouse on the Legitimation of Scientific Knowledge. *Philosophy of the Social Sciences, 33*(4), 444–463.
Remedios, F. (2009). Fuller and Mirowski on the Commercialization of Scientific Knowledge. In J. van Bouwel (Ed.), *The Social Sciences and Democracy*. Basingstoke: Palgrave Macmillan.
Remedios, F. (2013). Review of Humanity 2.0: What It Means to be Human Past, Present and Future. *LSE Review of Books*. http://blogs.lse.ac.uk/lsereviewofbooks/ 2013/05/07/book-review-humanity-2-0/
Remedios, F. (2015). Knowing Humanity in the Social World: A Social Epistemology Collective Vision? In J. Collier (Ed.), *The Future of Social Epistemology: A Collective Vision* (pp. 21–28). London: Rowman and Littlefield International.
Remedios, F. (2016). Steve Fuller: Knowledge, the Philosophical Quest in History. *Metascience, 25*(1), 3–23.
Remedios, F., & Dusek, V. (2016). Fuller's Social Epistemology and Epistemic Agency. In P. J. Reider (Ed.), *Social Epistemology and Epistemic Agency* (pp. 61–74). London: Rowman and Littlefield International.
Remedios, F., & Dusek, V. (2018). *Knowing Humanity in the Social World: The Path of Steve Fuller's Social Epistemology*. New York/Basingstoke: Palgrave Macmillan.
Rescher, N. (1989). *Cognitive Economy: Economic Perspectives in the Theory of Knowledge*. Pittsburgh: University Pittsburgh Press.
Reverby, S. M. (2009). *Examining Tuskegee: The Infamous Syphilis Study and Its Legacy*. Chapel Hill: University of North Carolina Press.
Rifkin, J. (1983). *Algeny*. New York: Viking Press.
Robinson, J. (Ed.). (1990). *The Nag Hammadi Library*. New York: HarperOne.
Roco, M. C., & Bainbridge, W. S. (2003). *Converging Technologies for Improving Human Performance*. Dordrecht: Kluwer.
Rose, S. (2011). Practicing Biochemistry Without a License. *EMBO Reports, 12*(5), 381.
Rosenberg, H. (1980). *Atomic Soldiers*. Boston: Beacon Press.
Roth, P. (1991). The Bureaucratic Turn: Weber Contra Hempel in Fuller's Social Epistemology. *Inquiry, 34*, 365–376.
Rouse, J. (1996). *Engaging Science: How to Understand Its Practices Philosophically*. Ithaca: Cornell University Press.
Rouse, J. (2002). *How Scientific Practices Matter: Reclaiming Philosophical Naturalism*. Chicago: University of Chicago Press.
Rouse, J. (2003). Remedios and Fuller on Normativity and Science. *Philosophy of the Social Sciences, 33*, 464–471.
Ruse, M. Personal Communication.

Schatenschneider, D. (2004). *Visions of Symmetry: M. C. Escher* (2nd ed.). New York: Harry M. Abrams.

Schmidt, J. C. (2011). Toward an Epistemology of Nano-technosciences: Probing Technoscience from a Historical Perspective: On Today's Surprising Prevalence and Relevance of Francis Bacon. *Poiesis & Praxis, 8*, 103–124. https://doi.org/10.1007/s10202-011-0104-z.

Schroedinger, E. (1944). *What Is Life? The Physical Aspect of the Living Cell.* Cambridge: Cambridge University Press.

Schweber, S. (1982). Demons, Angels, and Probability: Some Aspects of British Science in the Nineteenth Century. In A. Shimony & H. Feshbach (Eds.), *Physics as Natural Philosophy: Essays in Honor of Laszlo Tisza* (pp. 319–363). Cambridge: The MIT Press.

Seeger, R. J., & Cohen, R. S. (1974). *Philosophical Foundations of Science.* Boston: Reidel.

Segerstrale, U. (2000). *Defenders of the Truth, the Battle for Science in the Sociology Debate and Beyond.* Oxford: Oxford University Press.

Senechal, M., & Fleck, G. (1972). *Patterns of Symmetry.* Amherst: University of Massachusetts Press.

Shiffman, M. (2015, November). Humanity 4.5. *First Things*, pp. 23–30.

Shubnikov, V. I., & Kopstik, A. M. (1974). *Symmetry in Science and Art.* New York: Plenum Press.

Simon, H. (1981). *The Sciences of the Artificial.* Cambridge, MA: MIT Press.

Simon, H. (1991). Bounded Rationality and Organizational Learning. *Organization Science, 2*(1), 125–134.

Sivin, N. (1995). On the Limits of Empirical Knowledge in the Traditional Chinese Science. In *Science in Ancient China: Researches and Reflections.* Brookfield: Variorum.

Snyder, L. J. (2011). *The Philosophical Breakfast Club: Four Remarkable Friends Who Transformed Science and Changed the World.* New York: Broadway Books.

Snyderman, M., & Rothman, S. (1988). *The IQ Controversy: The Media and Public Policy.* New Brunswick: Transaction Publishers.

Sonne, A. (2015). Canaries, Germs, and Poison Gas: The Physiologist J. S. Haldane's Contributions to Public Health and Hygiene. *Dansk Medicinhist Arbog, 43*, 71–100.

Speier, F. (2015). *Big History and the Future of Humanity.* Chichester: Wiley.

Stamos, D. N. (2001). Quantum *Indeterminism* and Evolutionary Biology. *Philosophy of Science, 68*(2), 164–184.

Stoppard, T. (1994). *Arcadia.* New York: Farrar, Strauss and Girardoux.

Sustainable Knowledge: An Exchange, Davis, Dieleman, Frodeman, Remedios, R., Simbürger, S. https://social-epistemology.com/2014/03/15/sustainable-knowledge-an-exchange-davis-dieleman-remedios-riggio-simburger-suomela/#more-5189

Sytsma, J., & Buckwalter, W. (Eds.). (2016). *Companion to Experimental Philosophy*. Chichester: Wiley.

Taubes, G. (2002, April). Postol Versus the Pentagon. *Technology Review*. http://www.technologyreview.com/featuredstory/401412/postol-vs-the-pentagon/

Teillard de Chardin, P. (1959). *The Phenomenon of Man*. With an Introduction by Julian Huxley. New York: Harper.

Toulmin, S. (1963). *Foresight and Understanding*. New York: Harper Torchbooks.

Truesdell, C. (Ed.). (1960). The Rational Mechanics of Flexible or Elastic Bodies 1688–1788. *Leonardi Euleri. Opera Omnia. XI*, Part 2, 288.

Tversky, A., & Kahneman, D. (1974). Judgment Under Uncertainty: Heuristics and Biases. *Science, 185*, 1124–1131.

Tversky, A., & Kahneman, D. (1983). Extensional Versus Intuitive Reasoning: The Conjunction Fallacy in Probabilistic Reasoning. *Psychological Review, 90*, 293–415.

Tyson, et al. (2008). Biological Switches and Clocks. *Journal of the Royal Society Interface, 5*(Supplement 1), S1–S8.

Uebel, T. (2005). Review of Legitimizing Scientific Knowledge: An Introduction to Steve Fuller's Social Epistemology. *Notre Dame Philosophical Reviews*. http://ndpr.nd.edu/news/24412-legitimizing-scientific-knowledge-an-introduction-to-steve-fuller-s-social-epistemology/

Van Fraassen, B. (1991). *Quantum Mechanics: An Empricist View*. Oxford: Oxford University Press.

Vernadsky, V. I. (1944). Problems of Biogeochemistry II (trans. Vernadsky, G.). In G. Evelyn Hutchinson (Ed.) *Transactions of the Connecticut Academy of Arts and Sciences, 35*, 483–517.

Vernadsky, V. I. (1998). *The Biosphere*. Introduction by Lynn Margulis et al. New York: Copernicus.

Voegelin, H. (1954). *The New Science of Politics: An Introduction*. Chicago: University of Chicago Press.

von Goethe, W. (1967). *Theory of Colours* (trans. Eastlake, C. L.). London: Cass. (Orig. 1810).

von Humboldt, W. (1970). On the Spirit and the Organizational Framework of Intellectual Institutions in Berlin. *Minerva, 8*, 242–250.

Webb, E. (2005). Voegelin's Gnosticism Reconsidered. *Political Science Reviewer, 34*, 48–76.

Weinberg, S. (2007). *Facing Up: Science and Its Cultural Adversaries*. Cambridge, MA: Harvard University Press.

Weiner, J. (1999). *Time, Love, and Memory: A Great Biologist and His Quest for the Origins of Behavior*. New York: Alfred A. Knopf.

Wellsome, E. (1999). *The Plutonium File: America's Secret Medical Experiments in the Cold War*. New York: Dial Press.

Wennemann, D. (2013). *Posthuman Personhood*. Lanham: University Press of America.

Weyl, H. (1952[1989]). *Symmetry*. Princeton: Princeton University Press.

White, L. (1967). The Historical Roots of Our Ecological Crisis. *Science, 155*(3767), 1203–1207.
Wiener, N. (1964). *God and Golem Incorporated.* Cambridge, MA: The MIT Press.
Wingspread Statement on the Precautionary Principle. (1998). http://www.who.int/ifcs/documents/forums/forum5/wingspread.doc
Young, G. M. (2012). *The Russian Cosmists: The Esoteric Futurism of Nicholai Fedorov and His Followers.* Oxford: Oxford University Press.
Yoxen, E. (1983). *The Gene Business: Who Should Control Biotechnology?* New York: Harper and Row Publishers.
Zimbardo, P. (2008). *The Lucifer Effect: How Good People Turn Evil.* New York: Random House.
Zimmerman, N. H., & Menaker, M. (1979). The Pineal Gland: A Pacemaker Within the Circadian System of the House Sparrow. *Proceedings of the National Academy of Sciences USA, 76*(2), 999–1003.

Index[1]

[1] Note: Page numbers followed by "n" denote notes.

F.X. Remedios, V. Dusek, *Knowing Humanity in the Social World*,
https://doi.org/10.1057/978-1-137-37490-5

Kolibri 360 GmbH
Ferdinand-Jühlke-Straße 7
99095 Erfurt, Deutschland
produktsicherheit@kolibri360.de